THE WORLD OF SCIENCE
COMMUNICATIONS AND TRANSPORT

THE WORLD OF SCIENCE
COMMUNICATIONS AND TRANSPORT

MARK LAMBERT & JANE INSLEY

Facts On File Publications
New York, New York ● Bicester, England

COMMINUCATIONS AND TRANSPORT

Library of Congress Cataloging in Publication Data

Main entry under title:

World of Science

   Includes index.
   Summary: A twenty-five volume encyclopedia of
scientific subjects, designed for eight to twelve-year-
olds. One volume is entirely devoted to projects.
   1. Science—Dictionaries, Juvenile.   1. Science—
Dictionaries
Q121.J86 1984   500   84-1654

ISBN: 0-8160- 1073-0

Printed in Italy
10 9 8 7 6 5 4 3 2 1

Consultant editors
Eleanor Felder, Former Managing Editor, *New Book of Knowledge*
James Neujahr, Dean of the School of Education, City College
of New York
Ethan Signer, Professor of Biology, Massachusetts
Institute of Technology
J. Tuzo Wilson, Director General, Ontario Science Centre

Previous pages
High above the Earth,
the space shuttle is
photographed by a
communications
satellite at the start of
one of its missions.

Editor Penny Clarke
Designer Roger Kohn

CONTENTS

▶A member of the shuttle's crew is moved towards a work area by the arm of the orbiter's remote manipulation system, operated by another crew member.

Note There are some unusual words in this book. They are explained in the Glossary on pages 62–63. The first time a word is used it is printed in *italics*.

5

COMMUNICATIONS

USING LANGUAGE

The word 'communications' is used in many different ways. We speak of 'communications networks', 'communications satellites' and 'telecommunications'. But communication is basically very simple. When one person communicates with another, he or she passes on a piece of information, an idea or a feeling.

People have been communicating with each other for thousands of years. And the development of communications systems has had an enormous influence on the development of civilizations. In 3000 BC people living in Mesopotamia were not even aware that other parts of the world existed, let alone that there were other people. Today, as a result of advanced forms of transport and communications systems, people on opposite sides of the world can get in touch with each other very easily. In this way modern communications and transport have made the world seem smaller.

Speech and language
Man is not the only communicator. In fact most of the world's living creatures communicate in some way or other. In lower animals, such as some insects, communication is limited to attracting mates and warning off rivals. But some mammals have elaborate and subtle means of communication. For example, when a group of lionesses hunt together, each one seems to know what the others are doing. Dolphins appear to communicate by means of a simple language. And some people have even managed to communicate with chimpanzees, using a simple sign language.

Language is our most important means of communication. Early humans probably communicated mostly by signs, accompanied by grunts and growls. But by about 500,000 years ago they had developed the ability to speak. Now, instead of making sounds to indicate general ideas, such as 'danger' or 'food' people could say what form of danger and what kind of food. Gradually, they built up a vocabulary of words to describe the world about them. But people living in different places used different words, and so today there are many languages in the world.

When we speak we produce sounds, which are vibrations in the air. These air vibrations are produced by our vocal cords – two pieces of elastic tissue in the voice box, or larynx. Air coming up from the lungs makes them vibrate like the strings of a guitar. To produce high-pitched sounds muscles make the vocal cords tight. For lower sounds the muscles relax and the vocal cords slacken.

The sounds produced by the vocal cords are modified by changing the shape of other parts of the vocal tract – the throat, jaws, mouth cavity, tongue and lips. Using these we can produce vowel sounds (eg 'a', 'e', 'i', 'o', 'u') and consonants (eg 'b', 'f', 'g', 'm', 's') which can be put together to make words.

Writing and printing
People first began to record their ideas over 30,000 years ago by drawing pictures on the walls of caves. By 3,000 years ago picture languages had been developed. Special pictures, or

►Hieroglyphics carved and painted on the tomb of an ancient Egyptian nobleman.

◄In this ancient
Indian manuscript the
highly detailed
illustration is more
important than the
writing of the text.
This was also true of
many medieval
European manuscripts.

hieroglyphics, were used to represent not only objects but also ideas and feelings. The Egyptians wrote using hieroglyphics.

Meanwhile the Sumerians, who lived in the region where Iraq is today, had developed a different style of writing. They used patterns of wedge-shaped, or *cuneiform*, marks inscribed onto clay tablets. They recorded lists and descriptions of events in this way.

The first alphabet appeared in about 1200 BC in Syria. It contained 32 letters, each of which represented a particular sound. Today, most western countries use an alphabet of 26 letters. This first appeared in the *Middle Ages* and was derived from the Roman alphabet, which in turn came from the classical Greek alphabet first used around 400 BC.

As long as everything had to be carefully written out by hand, communication by writing was very slow. And it was possible to reach only a few people in this way. Indeed, very few people could write. Closely connected with this was the fact that few people could read. After all, learning to read serves no useful purpose if there is nothing to read.

But in about 1450 the German inventor Johann Gutenberg (1398–*c.*1468) devised the technique of printing with a movable type. Handwritten books had been rare and expensive. Printed books now became much cheaper and more easily available. Today vast amounts of information are printed every day in the form of books, magazines, newspapers, leaflets, advertizing posters, product wrappings and many other things. Information and ideas can now be communicated to many people very rapidly.

▲The angular letters of the Greek alphabet show very clearly in this stone carving. This was the first recognisable alphabet to develop and is the basis of the modern Greek alphabet.

►An elaborately illuminated, or decorated, medieval European manuscript.

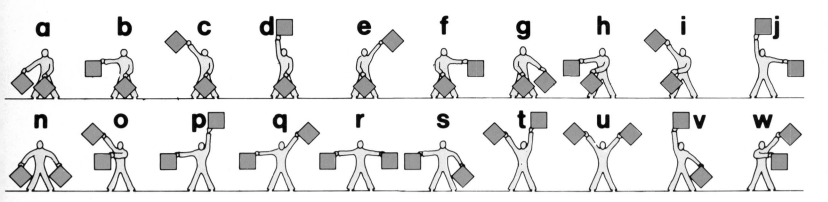

▲The letters of the alphabet in the internationally accepted semaphore code of signals.

▼The letters of the alphabet in the code devised in 1838 which is still used and still bears the name of its inventor: Samuel · Morse.

a	·—	n	—·
b	—···	o	———
c	—·—·	p	·——·
d	—··	q	——·—
e	·	r	·—·
f	··—·	s	···
g	——·	t	—
h	····	u	··—
i	··	v	···—
j	·———	w	·——
k	—·—	x	—··—
l	·—··	y	—·——
m	——	z	——··

Language on its own is not much use for communicating over long distances. Signals that could be seen were the first means of long-distance communication. Visual signals include such things as flashing lights, coloured lights and flags. Bonfires were the first type of visual signal to be used. In 1084 BC the Greek leader Agamemnon sent a message to his wife Clytemnestra telling her of the fall of Troy. He used a series of bonfires over a distance of 805 km (500 miles).

The semaphore telegraph

But bonfires could be used only to send prearranged messages. Other messages had to be carried by couriers, usually on horseback. During the 1790s a French engineer, Claude Chappe, devised a semaphore *telegraph*. This used long lines of semaphore towers. Each tower had two arms that could be moved to indicate different letters, enabling messages to be sent quickly from tower to tower. But this system had one big disadvantage – if the weather was bad the signals could not be seen.

The electric telegraph

The ancient Greeks knew about electricity 2000 years ago, but its power has only been understood and used for about the last 200 years. Early in the nineteenth century scientists were experimenting with it as a means of long-distance communication.

The first electric telegraphs used several wires to send a message. In 1837 Charles Wheatstone and William Cooke devised a telegraph that used five wires connected to needles. Different letters

were indicated by causing different combinations of needles to move. In 1838, however, Samuel Morse patented his single wire telegraph. To indicate different letters, he used his now famous Morse code of dots and dashes.

In Morse's system the dots and dashes in which the message was sent had to be decoded by someone at the receiving end of the wire and written out by hand. In 1855 Professor David Hughes, an English scientist working in America, invented a printing telegraph, or tele-typewriter. As the message arrived, it was automatically decoded and typed out by the machine. Modern telex and TWX machines work in a similar way.

The telephone

Telegraph communication became very important during the late nineteenth century. But in 1876 Alexander Graham Bell invented the *telephone* which turned human speech into electrical signals and transmitted them down a wire.

The American inventor Thomas Edison improved Bell's telephone by devising separate microphones and earpieces. By 1878 the first telephone exchange had been opened. But because the telegraph was by now well established, telephones did not become really popular until about 1920. The first transatlantic telephone service was introduced in 1927. This service used radio; the first transatlantic telephone cable was not laid until 1956.

Radio communications

The telegraph and telephone were useful for sending messages from one land station to another. But before the

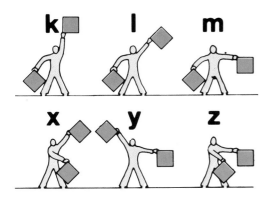

k   l   m

x   y   z

invention of wireless telegraphy, or radio, it was impossible to communicate with ships at sea. The existence of radio waves was known by the 1880s and in 1890 a French scientist, Edouard Branly, invented a device called a coherer, in which metal filings were used to detect radio waves. In 1894 the Italian scientist Guglielmo Marconi began to experiment with radio waves and in 1901 he transmitted the first radio signal across the Atlantic. In 1906 an American scientist, Reginald Fessenden, made the first-ever broadcast of speech and music. In the same year it was discovered that certain *crystals* were better at detecting radio waves than Branly's coherer with its metal filings. The first radio sets sold to the public were crystal sets.

▲A telephone exchange in 1882. The operator had to make all the connections by hand.

◄A very early telephone. You spoke into the hollow bell at the top. The earpiece hangs at the side ready for use.

►This telex will reach its destination within a few hours.

▼A crystal set – one of the earliest of radios. Today's 'walkmen' are its descendants.

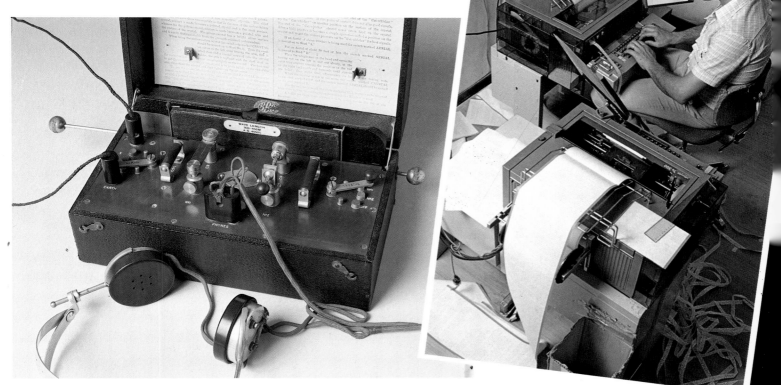

ELECTRONIC COMMUNICATIONS

Electronic devices are those that are used to control very accurately the flow of electric current. The first electronic device was the thermionic valve, or vacuum tube, invented by John Ambrose Fleming in 1904. Thermionic valves rapidly replaced crystals for detecting radio waves.

Valves were also the first electronic switches. They remained important components of electronic circuits until 1947, when the *transistor* was invented by three American scientists, Walter Shockley, John Bardeen and Walter Bratain. By 1957 factories were turning out 30 million transistors each year. By this time, as technology improved, transistors had become much smaller than when they were first invented. But in 1958 electronic components became even smaller. Integrated circuits, or silicon chips as they are usually called, were first made by Texas Instruments in the USA. It was now possible to place hundreds of transistors and other

components on a single tiny slice of silicon. These developments had far-reaching effects on communications systems.

Television

By the 1930s thermionic valves and other electronic devices had made it possible to transmit pictures as well as sounds by means of radio waves. Modern television is based on a device called the *cathode ray tube*, which was originally invented in 1897. The first television system was demonstrated in 1926 by the Scottish inventor John Logie Baird, who opened a television studio in London in 1929.

Baird used a mechanical system to produce his television pictures, but in 1931 Vladimir Zworykin produced the first electronic camera. Fully electronic colour television first came into use in 1953. And in 1960 the Japanese company Sony produced the first television in which transistors replaced the much less reliable electronic valves used before.

►An early television camera tube used for transmitting black-and-white television pictures.

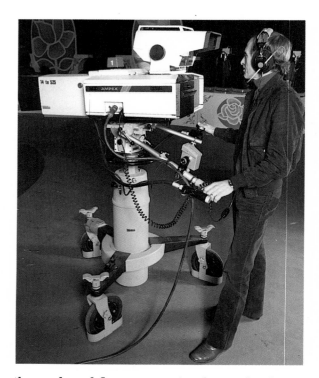

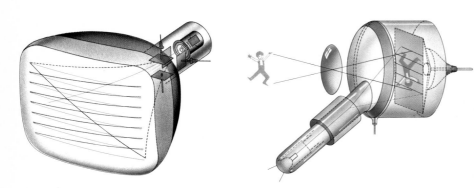

▲A modern black-and-white television camera tube (**above right**). In the camera (**far right**) light rays from the subject are focussed through a lens on to a light-sensitive screen and a target, setting up a pattern of electrons from an image of the subject. A narrow beam of electrons from an electron gun (projecting forward) then scans the target very fast, losing electrons where these are missing from the pattern. The electron beam now carries the electron pattern. It strikes a plate called the anode and flows out of the camera tube as an electric current or signal, which is transmitted via a cable or radio wave to a television receiver tube (**above**). Here, the signal controls a beam of electrons which scans a phosphor screen that glows as electrons hit it, so reproducing the subject on the television screen.

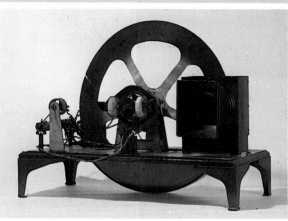

Three stages in the development of the television. **Far left** is the system invented by James Logie Baird. **Left** is a set produced in the 1940s and **below** is the latest streamlined portable model.

Communications satellites

Television signals cannot travel very far without being amplified, or boosted. And unlike radio waves, they cannot be bounced off the *ionosphere*. So for some time sending pictures round the world was impossible. But in 1958, one year after the first Russian Sputnik was launched, scientists in the USA began to experiment with communications satellites. In 1962 the first pictures were sent between Europe and the USA via the satellite 'Telstar 2'.

Today there are a number of communications satellites orbiting the Earth. The 'Intelsat' system of three satellites is typical. All three satellites have *geostationary orbits*, which means that they remain in the same places above the Earth as it spins. Each satellite covers over a third of the Earth's surface, so between them they cover the whole of the Earth. Together with modern cameras and other equipment, the 'Intelsat' system allows stories and pictures of news and sports events to be transmitted by journalists almost instantly to any part of the world.

▼A communications satellite is launched from the space shuttle in February 1984.

Telephones and computers

Electronic telephone exchanges were first introduced during the 1960s. At the heart of such exchanges were computers, which by then were beginning to develop rapidly as a result of the invention of the silicon chip. During the same period telephone companies began to experiment with satellites and in 1971 it became possible for telephone users in New York and London to dial each other directly. This was much quicker than the previous system when the person making the call rang the operator who then rang the person the caller wanted to speak to.

A modern communications satellite can provide over 6,000 telephone channels, as well as two television channels. Telephones themselves have also developed. A modern electronic telephone has a memory that can hold up to ten numbers.

Computers now play a large part in communications systems. Television companies broadcast information known as teletext. It comes from a central computer and can be displayed on the screens of televisions equipped with suitable decoders. Telephone companies produce a similar form of information service, known as viewdata, which is transmitted along telephone lines. The user's telephone is linked to the television and the information is displayed on the screen.

It is also possible to link two or more computers by telephone. Information typed on the keyboard of one computer is printed out on the printer of a receiving computer. Using a facsimile (fax) machine any kind of document can be sent by telephone – even photographs.

Electronic mail, which works in a similar way but just sends text, can be sent from one office to another far more quickly than letters can be sent through the postal system. Electronic mail can even be sent via communications satellites.

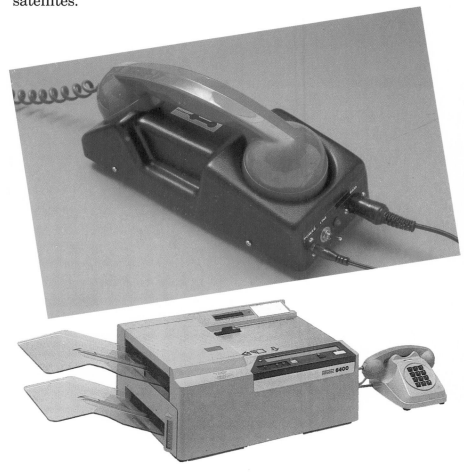

▼An acoustic coupler converts signals that the telephone cannot understand into sounds that it can transmit to the next part of a communications system.

◄A major future war may be controlled from computer terminals and television screens rather than the battlefield. This is how some military planners think such a control room would look.

▲Why bother with the mail when all you need do is get the right equipment and connect the telephone to it and wait for the messages to come through, printed out by the special printer. Many international firms are now using this form of communication.

2 TRANSPORT BY ROAD

ROADS AND ROADBUILDING

Transport and communications are closely linked. And sometimes they are really one and the same; written messages need to be carried by some means to the person they are addressed to. As early civilizations developed and towns grew up, the routes along which traders and messengers travelled became more and more important. Paths became wide tracks, but such well-used tracks rapidly deteriorated. In wet weather they must have been very muddy and almost impossible to use.

The first roads appeared as people began to dig drainage ditches along the sides of their trading routes. The soil from the ditches was transferred to the centre of the road, filling in hollows and levelling the roadway. Later, people began to surface their roads with hard materials. Stone-surfaced roads were in use in China over 3,500 years ago.

Roman roads

The Romans were the first scientific road builders. During the period of the Roman Empire over 85,000 km (53,000 miles) of roads were built. Roman roads are noted for their straightness. While other people built roads along old tracks that followed the easiest routes for horse and carts, the Romans preferred to follow straight lines. To do this they often had to cross marshes, ravines and mountains. The roads they built are still admired by modern engineers.

Like modern roads, a Roman road was built in layers. The foundation consisted of broken stones set in mortar. Above this was a layer of concrete made of small pieces of stone, sand and lime. The top layer was paving stones set in mortar.

European roadbuilders

After about AD 200, when the Roman and other major civilizations began to decline, roads fell into disrepair. It was not until the seventeenth century that people once again began to take an interest in roadbuilding. At this time stagecoaches were just starting to become popular. The roads they ran on were stone-surfaced and they were built in a similar way to Roman roads.

By the end of the eighteenth century the best roads were to be found in France. In 1775, the French engineer Pierre Trésuaget devised a new system of roadbuilding. His roads were based on the idea that the soil beneath the road should be made to take the weight of the traffic, rather than the surface layer of the road.

At the same time two Englishmen, John Metcalf and Thomas Telford were improving Britain's roads. Metcalf was particularly insistent upon good drainage and Telford's roads were built to take the heaviest of loads. The Scottish engineer, John McAdam used the ideas of all these engineers. He built roads from which rainwater drained rapidly. In 1816 he devised a new kind of surface which would keep the natural soil beneath the road dry. Macadamized road surfaces consisted of small stones laid in loose layers and compacted together by the traffic.

In Britain, Europe and the USA roadbuilding increased rapidly until about 1840. But then, as a result of the growth of the railways, roadbuilding virtually stopped until about 1910. But with the development of first the bicycle and then the motor car roadbuilding once again became essential.

▶In a modern road system major routes, such as motorways and freeways, are designed to keep traffic moving. Overpasses, underpasses and sliproads avoid the need for junctions where traffic can build up and cause delays.

electricity
telecommunications
gas
water

main sewer

sewer

asphalt
bitumen macadam

macadam base
shale sub-base

◀A cross-section of a street, showing the four layers that make up the roadway and the services that are laid in the ground below.

15

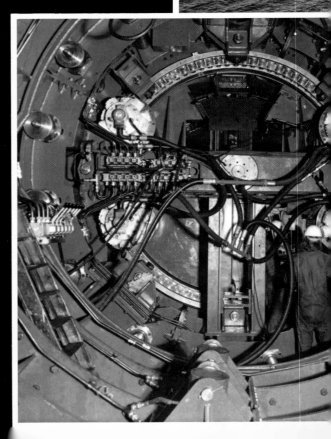

▶Roadbuilding often involves building bridges and tunnels. One of the world's greatest engineering feats is the Chesapeake Bay Bridge tunnel, which spans 27 km (17 miles) of water. There are two man-made islands which are linked to each other and the mainland by two bridges and two tunnels each 1 km (0.6 mile) long.

▼Inside the shield of one of the most modern pieces of tunnelling equipment. Roads in tunnels are becoming more important as the land surface becomes more expensive and more built on.

▲This four-decked drilling platform was used in the excavation of the Italian side of the road tunnel under Mont Blanc. The tunnel is 12 km (7½ miles) long. Engineering projects like this are very slow. The best rate of progress on this tunnel was about 9 m (29 ft) in a day.

Roads in the twentieth century

The development of the motor car meant that roads had to be built with much stronger surfaces. Today two main types of road surface are used. Minor roads that are used only by light traffic have surfaces of three layers. At the bottom is a sub-base course of large stones, compacted together by hammering. Above this is a well-rolled layer of ash or gravel. The surface is made of tarmacadam – small stones bound together with tar. The name of the surface is a reminder of the engineer who designed the first practical modern road surface.

A motorway or other road that is used by heavy traffic is much stronger. Above the sub-base course of stones is a layer of concrete, which may be reinforced with steel mesh. Above this lie a layer of waterproof paper and another layer of concrete. The surface may or may not be covered with tarmacadam.

WHEELS, CARTS AND CARRIAGES

◀Wheeled carts appeared in Europe in about 3000 BC. Shown here on the left is an early European solid wheel constructed from four pieces of wood carefully jointed together. On the right is a European spoked wheel dating from the early Iron Age (about 800 BC).

The wheel was one of the earliest and most important inventions made by humans. Its origins are a little uncertain, but it probably developed from the practice of moving heavy objects by rolling them on logs. The first wheels, however, were not used for transport – they were used to help make pottery. The potter's wheel appeared in Mesopotamia in about 3500 BC.

Solid wheels

We don't, of course, know when or who realized that the potter's wheel could have other uses, but wheels were certainly being used for transport by about 3200 BC. Such wheels were made from slices of wood, cut from tree trunks and fixed onto the ends of long poles, or *axles*. By 3000 BC the Sumerians were using solid wheels made from two or three pieces of wood clamped together with cross-struts. This meant that wheels could be made using planks cut from the hard wood at the middle of a tree, avoiding the outer soft wood which wore away very quickly.

At the same time people discovered that carts could turn corners more easily if the wheels were not fixed to the axles. So they began building carts and wagons with the axles fixed to the body. The wheels, held on by pegs, could turn independently.

Spoked wheels

Solid wheels were heavy. This did not matter very much for slow-moving forms of transport, such as farm wagons. But war chariots, which needed to move

swiftly, had to be as light as possible. The spoked wheel was invented in about 200 BC and was used very effectively by the Babylonians, Egyptians and Greeks for their war chariots. Wheels could have any number of spokes up to forty, although those with four spokes were the most common. Four-spoked wheels also sometimes had a religious significance because the sun god was believed to drive across the heavens in a chariot with wheels with four spokes.

Tyres

Wooden wheels tended to wear down very quickly. The Sumerians tried to solve this problem by covering the rims with leather. Later, people used wheels with tyres made of pieces of bronze or copper, or with nails driven in as studs.

But the problem was not really solved until the Celts invented the iron tyre in about 700 BC. A circle of metal was heated to make it expand, fitted over the wheel and allowed to cool. As it cooled it shrank and tightened the wheel, which became much stronger. This process was used throughout the world until the English manufacturer Thomas Hancock made the first rubber tyres in 1846. Then in 1888 a Scottish vet, John Dunlop, devised the forerunner of our modern air-filled or pneumatic ones.

From carts to coaches

The earliest wheeled vehicles were two-wheeled carts. Very soon there were two-wheeled war chariots, pulled by wild asses. The Sumerians also appear to have had four-wheeled ceremonial wagons.

These must have been difficult to manoeuvre round corners as both axles seem to have been fixed to the body of the wagon.

The Romans used two shafts instead of one central one, which meant that a light cart could be pulled by one horse instead of two. And their four-wheeled wagons had pivoted front axles so that they could turn corners easily. Roman transport included the **raeda**, which was a four-wheeled wagon used for long-distance travel and a **carruca dormitoria**, which was a type of coach. The **biga** was a light racing chariot and a **carrus triumphalis** was a ceremonial chariot used by returning heroes. The Romans even had mail-wagons.

Although roads declined during the Middle Ages, travel actually increased. Passenger wagons with an early form of suspension appeared at this time. A central passenger container known as a litter was suspended by chains or leather straps from four tall posts fixed to the wagon-base.

The first coaches appeared in Hungary during the sixteenth century. By the seventeenth century coaches were in use all over Europe. They were now small and light and were equipped with steel springs. The most successful coaches were the Berlins, which were used as passenger and mail coaches throughout the eighteenth and nineteenth centuries. The stagecoaches of the nineteenth century got their name from the fact that they made long journeys broken at staging posts, usually inns, along the route. They were very comfortable because they used *laminated spring suspensions*. At the same time the Americans were using the famous Concord stagecoaches.

Horsedrawn vehicles were by now very popular and there were many different types. Two-wheeled types included the chaise, gig and curricle. Four-wheeled carriages included the barouche, landau, phaeton and hackney carriage.

▲An Egyptian nobleman's chariot with light, six-spoked wheels.

▼The earliest Sumerian wheels were made from three pieces of wood held together with batons. The wheels of the cart shown on this Sumerian mosaic, dating from about 2100 BC, are made from only two pieces of wood with an axle at the centre.

The motor car was not invented. It evolved gradually as a result of a number of different inventions and developments. The first practical self-propelled road vehicle was a steam-powered gun tractor built by the French engineer Nicolas Cugnot in 1769. Over the next 100 years a number of people built steam carriages. But they never really became popular. The public thought they were frightening and dirty. And the rich businessmen who owned the fleets of horsedrawn carriages saw them as a threat and did their best to discourage their use. In 1865 a law was passed limiting their speed to 6 km/h (4 mph) in the country and 3 km/h (2 mph) in the town. Until 1896 a man had to walk in front carrying a red flag. Even so, some steam vehicles were still built. The most successful was the Stanley, one of which broke the world speed record in 1906 with a speed of 204 km/h (127 mph).

Internal combustion engines
Meanwhile, some people were working on entirely different kinds of engine. The idea of an *internal combustion engine* – that is, one in which the fuel is burned inside the engine itself – was not new. In 1678 the Dutch scientist Huygens had tried to design an engine that used gunpowder as a fuel, but he was unsuccessful. Early in the nineteenth century scientists turned their attention to gas. In 1807 Isaac de Rivaz, a Swiss inventor, built an exploding-gas engine. In 1826 Sam Brown, an English engineer, tested a mechanical carriage that was powered by a gas engine; it proceeded up Shooters Hill on the outskirts of London with a rapid series of loud bangs.

The first successful gas engine was designed by the Belgian engineer Jean Lenoir in 1860. In 1876 the German engineer Nicolas Otto began building engines based on a new principle – the four-stroke cycle. This is a cycle in which only one stroke in four produces power. The remaining three strokes draw in fuel, compress it and push out the waste gases that are left after the explosion that produces the power.

Meanwhile an Austrian engineer, Siegfried Marcus, was experimenting with a new fuel. Petroleum oil had been discovered in 1859 and petrol was by now being used as a cleaning fluid. In 1865 Marcus built his first petrol-driven vehicle. He continued to build cars until the 1870s.

The next major advance was made by the German engineer Gottlieb Daimler, who had been Otto's production manager. In 1883 he built a high-speed, four-stroke engine that ran on petrol. He tried this engine out first on a bicycle and then, in 1886, on a four-wheeled car. At the same time another German engineer, Karl Benz, was also producing engines. And he began to see the commercial possibilities for motor cars. He built his first car in 1885 and in 1888 he became the first person to sell cars to the public.

Development of the car
Within a short time several manufacturers were producing motor cars. Among the first were two Frenchmen, René Panhard and Emile Levassor. They were granted the rights to make Daimler engines in France and they opened their factory in 1889. The first Panhard had its engine underneath the driver, which was the usual place at the time. (Now of course, the engine is generally under the vehicle's bonnet.) The 1891 Panhard, however, had a front-mounted engine. This position was soon adopted by all car manufacturers. The 1894 Panhard had a bonnet over the engine to keep out the weather and dirt. It also had a clutch and a set of four gears.

Other parts of the motor car that we take for granted today were added gradually by different manufacturers. Daimler's engine used a simple carburettor to mix the air and petrol (page 22). It was improved in 1888 when Edward Butler, an English engineer, devised the jet carburettor, which was improved still further in 1892 by the addition of the float chamber.

The battery and coil ignition was first

used by Karl Benz in 1885. The steering wheel was introduced in 1894 and in 1895 the Michelin brothers were the first to use pneumatic tyres on cars. By the early 1900s cars had such things as the honeycomb radiator for cooling, shaft drive from the engine to the wheels instead of chain drive, universal joints to allow movement of the shaft, accelerator pedal and drum brakes. Automatic transmission, electric starting, electric lights, dynamos and hydraulic brakes also date from this time.

Motoring through the years

Early in the nineteenth century three distinct types of car began to appear. First there was the small, relatively cheap family runabout. At the other end of the scale was the luxury car – expensive but beautiful. Lastly, there were the sports and racing cars.

Family cars appeared in the 1890s. De Dion Bouton was a French company that made small engines. These were put into light cars with bicycle-type spoked wheels. De Dion Bouton cars were copied by a number of manufacturers. The 'Baby' Renault was a particularly popular make.

But the big breakthrough in family motoring came in 1908 when Henry Ford introduced the Model T Ford. This was the first mass-produced car and between 1908 and 1927 over 15 million were built. Family cars that followed included the Morris Cowley and the 'Baby' Austin 7 in the 1920s and the later Volkswagon 'Beetle' and the British Leyland Mini. The VW 'Beetle' overtook the Model T's record in the early 1970s – so far over 20 million have been sold.

Early in the twentieth century the Daimler Company decided to build a safe, fast car. They produced the first Mercedes, named after Mercedes Jellinek, the daughter of the Austrian Consul in the French city of Nice, who had done much to help the company find rich clients. The first Mercedes was a very expensive sports tourer, but it set the pattern for all kinds of later motor vehicles. The 1907 Rolls Royce 'Silver Ghost' was very soon regarded as the peak of motoring luxury. Other cars of this type included the 1905 Peugeot, the 1907 Benz and the 1911 Austin Landaulet.

The luxury-car business was badly hit by World War 1 (1914–18) and the world-wide slump in trade that followed and during the 1920s many companies went out of business. Today, a small luxury market remains. But even family cars now have a great deal of sophistication and comfort.

The first motor race took place in France, between Paris and Bourdeaux in 1885. The winning car was a Panhard and Levassor driven by Emile Levassor himself. During the early years of the twentieth century speed was one of the main things that spurred manufacturers into producing better cars. At first the racing cars looked very much the same as cars driven on the roads, but by the 1920s specialized racing cars had appeared.

The modern car

A car consists of several different systems. The engine and body are supported by the chassis, which is attached to the wheels via the suspension system. This is needed to provide a comfortable ride and to keep the wheels on the road.

The engine is supplied with petrol or gasoline from the fuel tank via the carburettor, which mixes the fuel vapour with air. The fuel/air mixture is burned explosively in the engine and the waste gases are removed via the exhaust system. The moving parts of the engine are kept lubricated with oil and the engine is cooled either by air or by water that flows through a radiator.

The speed of the engine is controlled by a throttle in the carburettor, operated by a cable linked to the accelerator foot pedal. The power from the engine is transmitted to the wheels via a clutch, gearbox and transmission system. The clutch is a device for connecting and disconnecting the engine from the gearbox, which is used to reduce or increase the speed at which the wheels rotate in relation to the speed of the engine. The transmission consists of shafts and movable joints that carry the power to the wheels. Each wheel is equipped with a brake that is operated by

◄The thrill and speed of motor racing is vividly captured in this photograph taken during an American championship race. But today's motor racing is very much a team effort. The mechanics in the pits (**above**) must go through the routine checks fast and efficiently, at the same time looking out for and responding swiftly to unforeseen emergencies, because it's not just the drivers who keep the cars on the roads. Modern racing cars are so unlike anything you see on the streets that it's hard to realize that motor racing started soon after the first cars appeared and has been responsible for many advances in car design and engineering. Pneumatic tyres, for example, were first used on racing cars.

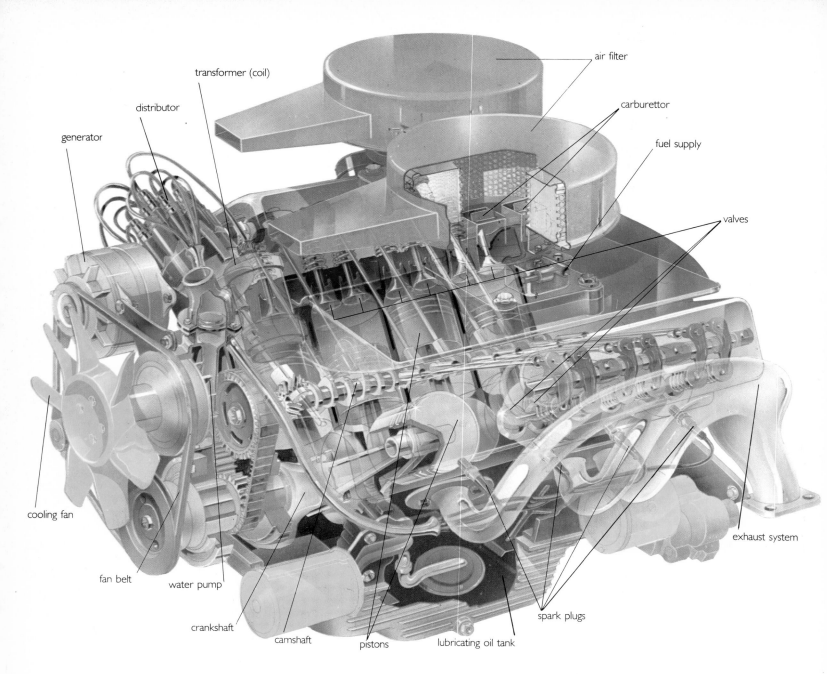

transformer (coil)

distributor

generator

air filter

carburettor

fuel supply

valves

cooling fan

fan belt

water pump

crankshaft

camshaft

pistons

lubricating oil tank

spark plugs

exhaust system

▲A V-8 petrol engine. Air is drawn in through two filters to the carburettors, which are supplied with fuel at the same time. Each cylinder has an inlet valve, which opens to allow fuel/air mixture to be drawn in. The valve closes and the fuel is ignited by a spark produced by the generator and coil and delivered to the spark plug via the distributor. The expanding gases drive the piston down the cylinder. As the piston rises again, the exhaust valve opens and the waste gases are removed via the exhaust system. The up and down movement of the pistons rotates the crankshaft, which drives the car wheels via the clutch, gear box and transmission. At the same time the crankshaft drives the generator, distributor, camshaft (which operates the valves), the water pump and the fan (which draws air through the radiator). Water from the radiator is circulated round the engine to cool it, and oil is pumped round from the sump to all the moving parts to reduce friction.

these components ensure that a spark is delivered to the fuel/air mixture in each engine cylinder at exactly the right moment. Other electrical components include the starter motor and lights.

Other types of car engine
Other types of engine can be used in cars. The diesel engine, invented in 1893 by Rudolph Diesel, was first used in road transport during the 1920s. Diesel engines are still used in such vehicles as lorries (which appeared on the roads soon after the first cars), taxis and all-terrain four-wheel drive vehicles. Today even some motor cars are powered by diesel engines. But diesel engines remain more expensive and noisier than petrol or gasoline engines.

Electric motors were first used in road vehicles as long ago as the 1890s. And in

hydraulic (fluid) pressure controlled by a foot pedal. The direction in which the front wheels point is controlled by the steering system.

Finally, the main electrical system of a car consists of a battery, dynamo, coil, distributor and spark plugs. Together

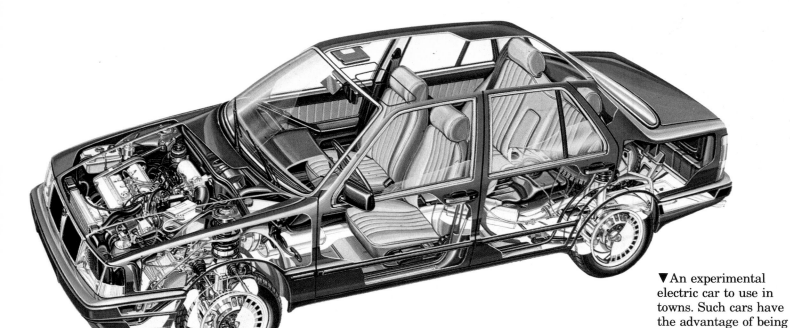

▲Cut away diagram of a modern four-door saloon car. It has four-wheel independent suspension and disc brakes on all wheels. This braking system has sensors at each wheel which are connected to an electronic control unit and an electro-hydraulic unit inside the engine bay. Working together, these elements vary the pressure on the discs so that brakes do not lock and cause the car to skid.

▼An experimental electric car to use in towns. Such cars have the advantage of being quiet and clean.

▼A car with a gas turbine engine built by the Rover Company of Great Britain. Rover's first gas turbine car could travel at about 245 km/h (152 mph). The main problem with gas turbine cars is that they are slow to accelerate and provide little or no engine braking when the throttle is released. Even so several companies are putting a great deal of time and money into gas turbine research.

1899 Camille Jenatzy, a Belgian driver, broke the land speed record with a speed of 106 km/h (66 mph) in an electric car. However, storing sufficient power in batteries has always been the main problem with electric cars. So although there are electric vehicles, such as milk floats and even a few prototype cars, the motor industry awaits the development of a powerful but lightweight battery. When this happens the cleanness of such vehicles will make them very popular.

Some manufacturers have also experimented with gas turbine engines. Again the main problems have been expense and weight. At the moment it seems more likely that gas turbine engines will be used in heavy vehicles, such as lorries and buses, rather than cars.

THE AGE OF STEAM

At the beginning of the eighteenth century Britain was in the middle of the *Industrial Revolution*. An efficient means of transport was urgently needed. The different components of railways already existed, all that was needed was people with enough vision to put them together.

Rails and horses

The idea of moving trucks on rails dates from the sixteenth century. Miners used two rows of planks to raise their trucks off the uneven floors of underground tunnels. Soon the planks were replaced by rails and some mines had wagons with flanged wheels, similar to those on a modern railcar, to keep them on the rails.

During the eighteenth century iron rails were introduced. At the same time mine railways were extended above ground to local wharves on rivers and canals and before long horses were being used to pull trains of wagons. The first horse-drawn passenger railway opened in 1807 between Oystermouth and Swansea. Horse-drawn trains continued in use for a number of years, particularly in France and America.

The early days of steam

Steam-powered vehicles appeared first on the roads. Among the pioneers of steam transport was the English inventor Richard Trevithick. He built several steam carriages, but found that people had little interest in them and so he turned his attention to railways. He developed his first steam locomotive in 1803 and in 1808 he was demonstrating his fourth locomotive in London. Called 'Catch-me-who-can' it pulled a single passenger carriage round a small circular track. Trevithick charged passengers 1 shilling (5p) for a ride (although 1 shilling then was worth far more than 5p or even 5 cents today).

Trevithick was plagued by bad luck and lack of money and he gave up building locomotives. But others took up the idea and by the 1820s a number of coal mines were equipped with steam-driven locomotives that ran on rails.

In 1814 the English inventor George Stephenson built his first locomotive, the 'Blücher', which was used at Killingworth

▼The scene at Nuremberg in December 1835 as George Stephenson's engine, 'Der Adler', gets up steam to haul the first train from the station. The covered carriage looks very similar to horse-drawn carriages.

▶Stephenson's 'Rocket', the winner of the 1829 Rainhill Trials, was the design on which all later locomotives were based. It can be seen in the Science Museum in London. Unfortunately, this famous locomotive is incomplete. The double-walled shell of the firebox is missing (only the back plate has survived). As a result the openings of the boiler flues are exposed.

colliery. In 1821 Stephenson was appointed chief engineer of the Stockton and Darlington railway. Built to carry coal from Durham to the coast, it was opened in 1825 with a train of passenger wagons, pulled by 'Locomotion', Stephenson's latest locomotive.

The most famous of Stephenson's many locomotives was the 'Rocket'. This was a new design of locomotive that he built for the Rainhill Trials, a competition for locomotives, in 1829. The 'Rocket' won the trials and, as a result, the Liverpool and Manchester Railway was equipped with 'Rocket'-type locomotives when it opened in 1830. This was the first railway to begin a regular passenger service and its opening marked the start of the railway age.

Railway networks soon began to spread, not only in Britain but also in France, other European countries and the USA. During the next 90 years over 1,100,000 km (nearly 700,000 miles) of track were built throughout the world.

Steam locomotives

A steam locomotive works in a very simple way. Fuel is burned to heat water and generate steam which is then piped to a cylinder containing a piston. Valves operate to pass the steam first to one side and then to the other side of the piston. Rods and linkages translate the to and fro movement of the piston into the rotational movement of the locomotive's driving wheels.

The steam locomotive was gradually refined over the years. Trevithick discovered that exhaust steam passing out of the chimney could be used to help draw the fire through the boiler, thus helping to raise the water temperature quickly. In 1827 Henry Booth, an English engineer, patented the multi-tubular boiler, in which the fire was drawn through many tubes instead of just one flue. It was this type of boiler that Stephenson used in the 'Rocket'.

Early locomotives had vertical cylinders and some had vertical boilers. The 'Rocket' had inclined cylinders, but by 1850 all locomotives had horizontal cylinders. Although during the next 100 years many different types of locomotive were built – different tasks called for different designs and there was a constant drive to improve efficiency – from 1830 onwards the 'Rocket'-type locomotive was the basic design, an indication of the genius of its designer. There were modifications of course, and these included longer boilers, various numbers and arrangements of wheels and the provision of a cab for the driver.

▼The 'George Washington', an early American locomotive, built by William Norris in 1836. This was the first 4-2-0 design, with the firebox behind the driving axles.

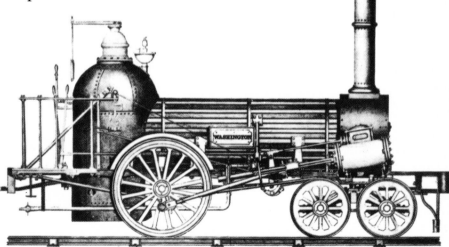

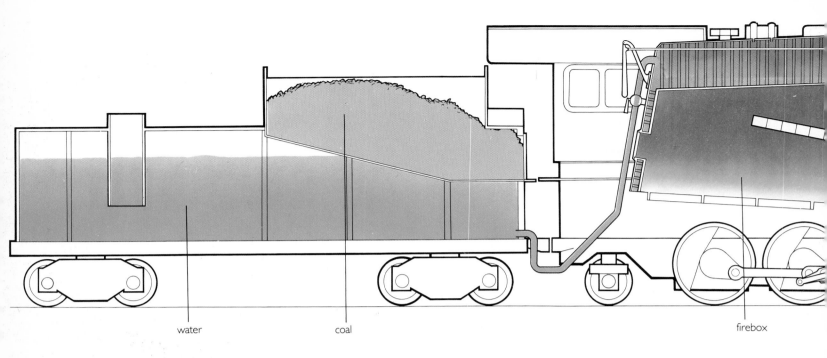

water   coal   firebox

▲Diagram of a steam locomotive. Coal is burned in the firebox and the hot gases pass through the boiler flues. This heats the water and produces steam above the water, in the space at the top of the boiler. The steam collects in the dome and is fed into a pipe via the main steam valve, which is controlled by the driver by means of a regulator handle. The steam passes back into the boiler in superheating pipes, each of which passes several times along a boiler flue. The superheated steam is passed to the valves and cylinders, which drive the wheels. The expanded steam leaves via the blast pipe and the chimney. As it leaves it helps to draw the fire through the boiler flues.

▶A British Railways Britannia class 4-6-2 locomotive. These locomotives were introduced in 1951, after British Railways had been nationalized, and were used to haul express passenger trains during the last days of steam.

In America most locomotives burned wood. Their huge smoke-stacks were designed to prevent sparks flying out. British locomotives used coal and coke as fuels. Coal-burning locomotives produced clouds of dirty, black smoke. Then, in 1860, someone (sadly we don't know who) had the idea of putting a brick arch in the firebox. This had three advantages. First, the flue pipes were protected from the direct heat of the fire. Second, the flue gases were forced to pass over the fire before entering the flue. Unburned gases and coal particles thus had an additional chance to burn properly and so the coal was burned more efficiently. Last, because of the increase in burning efficiency, the amount of smoke and cinders expelled from the chimney was greatly reduced.

Steam locomotives are described by three or four numbers that indicate the number and arrangement of their driving wheels and carrying wheels. For example

a 4-4-2 locomotive has four front carrying wheels, four driving wheels and two rear carrying wheels.

During the late nineteenth and early twentieth centuries some very large locomotives were built. Steel was by now available and the new steel boilers could withstand much greater pressures than the earlier iron ones. At the same time the German inventor Dr Wilhelm Schmidt devised a system of producing superheated steam, which enabled locomotives both to haul greater loads and use fuel more economically. The largest locomotives were the huge *articulated* Mallets and Garratts built in the US during the early 1900s. They were designed for pulling loads up steep inclines. The largest locomotive ever built was a Mallet – the Union Pacific Railroad's 'Big Boy' 4-8-8-4, which weighed 350 tonnes (389 tons).

In spite of the enormous variety of locomotives built during the early years of the twentieth century, the end of the age of steam was in sight by the 1940s. Road transport was increasing rapidly and railways were beginning to decline. At the same time diesel and electric locomotives were beginning to take over from the steam ones. Steam locomotives can still be seen in many countries where they have been preserved by enthusiasts or for tourism. Some still operate commercially in India, Africa and South America.

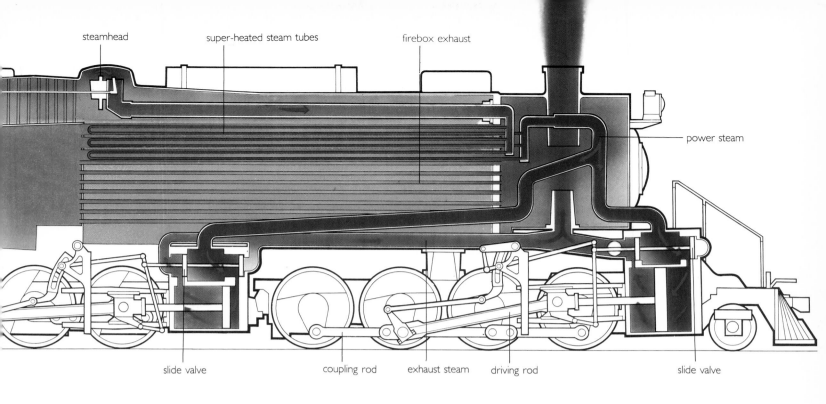

steamhead super-heated steam tubes firebox exhaust

power steam

slide valve coupling rod exhaust steam driving rod slide valve

MODERN TRAINS

Electric railways

In an electric railway electricity is supplied to the locomotives via overhead cables or electrified rails. The first electric rail vehicles were built during the 1830s, but they carried their own heavy batteries, which had to be recharged. It was not until 1883 that an engineer called Magnus Volk built the first proper electric railway along the seafront at Brighton. This railway still exists. Electric current is supplied from a third rail.

The overhead conducting wire was devised by a Belgian-American called Van de Poele in 1885. In 1888 another American, Frank Sprague, invented the swivel trolley pole, which led to the modern *pantograph* that collects current from overhead wires. Frank Sprague also invented the multiple-unit train, which is now a very common type of electric train. One driver can control a whole train of motorized coaches from either end of the train. Multiple-unit electric trains are most often found on underground railways.

From the late 1890s onwards many

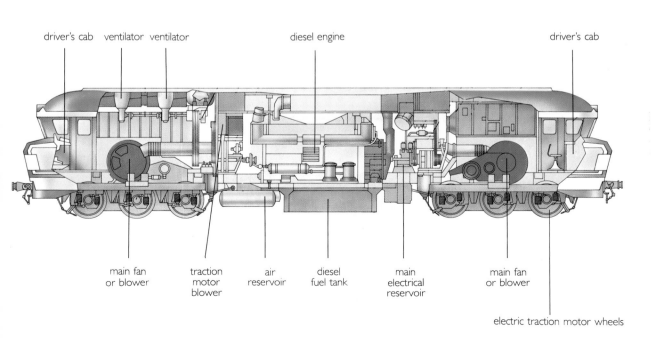

driver's cab ventilator ventilator diesel engine driver's cab

◀The diesel engine turns an electric generator which provides electrical power to the traction motors connected to the wheels which actually move the locomotive along.

main fan or blower | traction motor blower | air reservoir | diesel fuel tank | main electrical reservoir | main fan or blower

electric traction motor wheels

different types of electric locomotive were built. Modern locomotives are described by a code that indicates the arrangement of the axles. Carrying axles are indicated by numbers. The driving axles are indicated by letters (A=1, B=2, C=3, D=4, and so on) and a letter followed by 'o' indicates that the driving axles are not coupled together but are driven by separate motors. A popular modern arrangement is Bo-Bo (that is, two pairs of driving axles, each of which are individually driven), which appeared first on the Baltimore and Ohio railway in 1894.

Electric trains have several advantages. They are clean, have good acceleration and can climb steep slopes more easily than steam or diesel locomotives. However, electric railways are expensive to build and to operate.

Diesel-electric locomotives

Although we often speak of diesel locomotives, it is more accurate to call them diesel-electric locomotives. The diesel engine does not itself drive the wheels. Instead it drives a generator, which produces electricity to drive electric motors and these drive the wheels. Excess electricity is stored in batteries.

The first diesel-electric locomotive appeared in 1924. Typically, modern diesel-electric locomotives are mounted on

two *bogies* with either a Bo-Bo or Co-Co arrangement. A diesel-electric locomotive works more efficiently than a steam locomotive, but is more expensive to build and cannot pull such heavy loads. Like electric locomotives, they are often used in multiple units. In America as many as five units may pull a very heavy train.

High speed trains

In some parts of the world railway tracks have been altered or even specially built to take high speed trains. Among the fastest of these are the Japanese 'bullet trains' that run between Tokyo and Osaka. They are streamlined multiple-unit electric trains that travel at an almost constant 210 km/h (130 mph). In Britain high speed diesel-electric trains achieve speeds of 200 km/h (125 mph). In France some high speed trains are powered by gas turbine engines. In test runs such trains have reached 300 km/h (186 mph). The British Advanced Passenger Train (APT) is also powered by a gas turbine.

Special railways

Unusual kinds of railway include monorails and mountain railways. People began experimenting with the idea of using one rail instead of two as long ago

▼Two bullet-nosed Japanese Hikari, or 'lightning trains' at Tokyo station on the Shinkansen line. These huge multiple-unit electric trains, capable of seating 1400 passengers, leave every quarter of an hour.

◄The 'People Mover', a fully automatic driverless train developed in the USA for transporting large numbers of people between airport terminals.

as the 1820s. But the first true monorails were not built until the early 1900s. Between 1903 and 1910 several monorails were built in which gyroscopes were used to keep the trains upright. Today, however, monorail trains are either suspended from an overhead rail or sit on a concrete beam. Trains of the latter type run on rubber covered wheels and are kept upright by horizontal wheels that grip the sides of the beam.

The first mountain railways were built in the 1860s. James Fell designed the railway over the Mont Cenis Pass between Italy and France which was opened in 1868. To help the locomotives get uphill they had a pair of horizontal wheels that gripped a central rail.

A number of Fell railways were built, but the most successful type of mountain railway proved to be the rack or cog railway, in which a cog on the locomotive meshes with an extra toothed rail. The first, and perhaps best-known, of these railways was built in the US by Sylvester Marsh as a tourist attraction on Mount Washington.

▼The railway of the future? This small maglev (magnetic levitation) train is part of a shuttle service carrying passengers between Birmingham airport and the National Exhibition Centre just outside the city. Larger and more powerful versions of this monorail train have enormous potential as passenger transport (see page 61).

RAILWAY ENGINEERING

Thanks to the computer controlling trains is easier than ever before. The signalling centre at London's Victoria Station (**right**) is linked to another near Brighton and together they control 882 km (551 miles) of track. The diagram (**right**) shows a typical station on the line they control in the rush hour. The fast lines are shown in red, the slow ones in blue. Green indicates track that can be used by trains going in any direction. The dark green blocks represent the platforms. At peak times trains run through here at 2-minute intervals. Computers at the signalling centre monitor the positions of the trains, and the signals are altered to direct the trains along whichever track or to whichever platform is free.

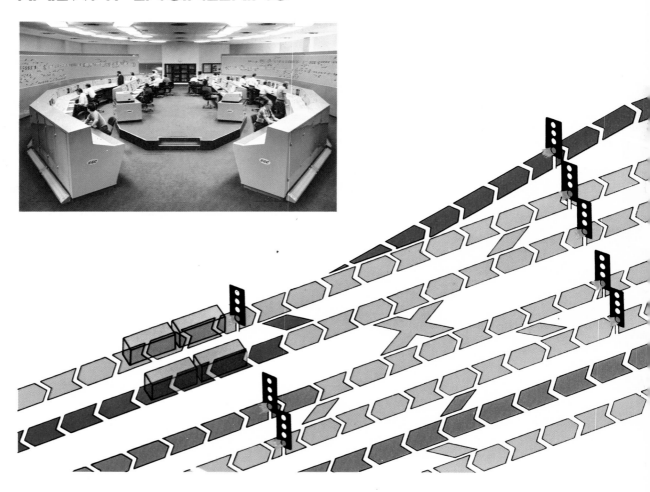

Tracks and signals

Railway track has to be laid very precisely on firm ground. The first stage is to level the roadbed, as it is called, and compact it well down. Then a layer of sand is laid on it, followed by a thick layer of ballast – sharp stones. On top of this the sleepers are laid. On older track these are wooden, but today concrete sleepers are usually used instead.

Fixed to each sleeper are two base-plates. The rails are held in place by fixing plates that bolt on to the base plates. Rails used to be laid in 30-m (100-ft) lengths. But today continuous welded rail is mostly used instead. This type of rail is laid in lengths of up to 300 m (980 ft) by special trains. The lengths are then welded together on the spot.

The first signals were used on railways in 1889. They consisted of vertical poles to which were fixed one or two movable arms. From the position of the arms, the engine driver knew if it was safe to proceed or not. In 1920 coloured lights

were introduced and these have now largely replaced the old semaphore signals. Track circuits – electric circuits connected to the rails – operate the signals automatically so that trains do not run into each other.

The track circuits are also linked to electronic, illuminated track diagrams in control rooms, which show the location of all the trains in an area. Using push-buttons, controllers can operate motorized points and direct trains along the correct routes.

Electronics is also playing an increasing role in marshalling yards, where freight trains are made up. The wagons are shunted into a yard via a hump. From there they roll down into their sidings. The controller selects the required siding on a control panel and the computer guides the truck forward, at the same time causing wagon retarders attached to the tracks to provide just the right amount of braking.

In some yards even the human

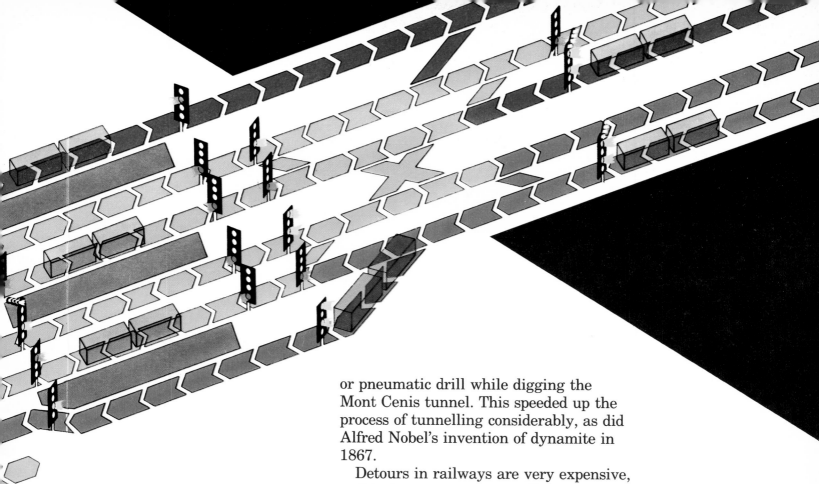

controllers have been replaced. The trucks are marked with computer bar codes and the computer decides which siding to send them to as they come over the hump.

Bridges and tunnels

Most railways need to be as level and as straight as possible. Often this involves cutting through mountains and hillsides or spanning valleys and rivers.

The shallowest tunnels are built by a method called 'cut and cover'. A deep wide trench is dug and then roofed in. Deeper tunnels have to be bored. In the early days of railways gunpowder and hand tools were all that was available, and tunnelling was a long and difficult task.

The tunnelling shield was invented by Marc Isambard Brunel in 1818 and this machine made it possible to tunnel through soft ground. As the shield moved forward, cutting through the ground, the soil was passed back on a conveyor belt and the tunnel behind was lined with bricks. Today ready-made concrete sections are used instead of bricks.

Tunnelling through rock remained a problem until 1861, when Germain Sommellier invented the compressed air or pneumatic drill while digging the Mont Cenis tunnel. This speeded up the process of tunnelling considerably, as did Alfred Nobel's invention of dynamite in 1867.

Detours in railways are very expensive, and trains are very heavy. So engineers had to build strong bridges. It was soon discovered that *beam bridges* were too weak over long spans. An early solution was devised by Robert Stephenson, George Stephenson's son, who built an iron *box girder bridge* over the Menai Straits in North Wales. But the most massive railway bridges are the great *cantilever bridges*, such as the Forth Railway Bridge in Scotland.

▼Mountainous regions provide a great challenge to railway engineers. The stone-built Weissen viaduct carries the narrow-gauge Rhaetian railway across an alpine valley in Switzerland.

SAILING SHIPS AND BOATS

Since the very earliest days of civilization rivers and other waterways have proved convenient for transport and communication. The first boats were simple dugout canoes, made by hollowing out tree trunks. Another type of canoe was made by stretching animal skins over a framework. In Mesopotamia wood was scarce. Reeds, however, were plentiful and boats were made by tying bundles of reeds together.

The ancient Egyptians built boats using papyrus reeds. But they also began to construct wooden sailing boats for use on the River Nile. These were built out of hundreds of pieces of wood carefully jointed together (there were no nails). Some were over 40 m (130 ft) long.

As trade developed, ships began to venture onto the sea. At first they kept to coastal waters, using oars when the wind was blowing in the wrong direction. In Egyptian ships the oarsmen sat in two rows below the main deck. Eventually ships began to venture farther from land and sailors had to learn the art of navigation.

▼A modern reconstruction of a papyrus boat of the type that the Egyptians sailed on the River Nile.

The Mediterranean sailors

The Phoenicians were probably the greatest of all the Mediterranean sailors. By 1200 BC they had perfected a very strong boat. Using cedar trees from Lebanon they built short, wide boats, no more than 20 m (65 ft) long. These boats were built of long planks strengthened with wooden ribs and heavy cross-beams. They were much sturdier than the Egyptian boats and stood up to rough weather better.

The Phoenicians established trading posts all round the Mediterranean Sea and even ventured beyond Gibraltar, reaching Cornwall in about 1000 BC. There they traded with the tin miners. One group of Phoenicians also seems to have made the first voyage around Africa – a journey that took them three years, as they had to stop at intervals to grow and harvest new provisions.

The Phoenician 'round ships', as they are sometimes called, were copied by both the Greeks and Romans. The Roman ships were larger – up to 30 m (100 ft) long – and their hulls had a lead covering to protect them from the devastating attacks of ship-worms, or piddocks, that bore easily through wood. The Phoenicians also pioneered the fighting galley, which was manned by soldiers, rowed by slaves and had an underwater ram at the bow. The Greeks and Romans copied these ships, too.

Viking longships and Chinese junks

As the Roman Empire came to an end, another type of boat was being developed farther north. For some time the only craft available to the northern peoples had been small boats, such as the Welsh coracle and Irish curragh. But between AD 500 and 700 the Europeans began building long, wooden boats, mostly for warlike purposes. The best and the most feared of the northern boatbuilders were the Vikings. In their longships they terrorized communities along the coasts of Britain and Europe. They also made long journeys and it is believed that the Vikings reached America 700 years

►Although damaged, this carving, which dates from about 700 BC, shows details of a Phoenician warship quite clearly. The soldiers with their round shields stand on the deck, while the slaves with their oars sit below. The underwater ram is very obvious.

▼Roman galley depicted on a coin from the 2nd century BC.

▼A life-size replica of a Viking longship. Viking warriors used such ships to sail along the coasts of northern European countries and terrorize the local people. They rarely sailed far from shore, although it is thought that Lief Am

▲A sampan on the Pearl River in China. This type of boat has changed little over the centuries, showing that it met local needs and sea conditions very efficiently.

The great age of exploration

The age of exploration was sparked off by the need to find a sea route to India which was the source of valuable spices, perfumes and silks. The land route previously used by merchants and traders had been cut off by the Ottoman Turkish empire since about 1300.

The main problem facing the explorers was navigation. Charts were either non-existent or highly inaccurate. Compass bearings were difficult to take accurately – sailors knew little or nothing about variations in the Earth's magnetic field. Finding a ship's position was also very hard. There were instruments for determining *latitude* and a good sailor could find his latitude to within 32 km (20 miles), but it was not until 300 years later that sailors had clocks accurate enough for them to work out their *longitude*.

In 1420 Prince Henry of Portugal (who was nicknamed Henry the Navigator) set up a school of navigation and seamanship at Sagres. He gathered together astronomers, mathematicians, ships masters, instrument makers and shipbuilders. Prince Henry sent out many exploratory expeditions. In 1487 Batholomew Diaz sailed round the southern tip of South Africa – the first European to do so since the Phoenicians. His crew forced him to turn back, but in 1498 the Portuguese explorer, Vasco da Gama, succeeded in reaching India by this route.

before Columbus.

Meanwhile in China another type of boat was being built. The Chinese were the first people to discover how to use the magnetic compass in navigation and their junks had three other important features. First, they had hinged rudders fixed to the sternpost – this replaced the inefficient steering oar used on all previous ships. Second, they used several masts instead of one and the extra sails gave them greater power. Third, the Chinese introduced watertight bulkheads, which meant that their ships did not necessarily sink if the hull was holed because the water was contained in one quite small place and the whole ship was not flooded.

The Spanish, too, sent out expeditions, and the typical ship used by both Spanish and Portuguese explorers in the fifteenth century was the caravel, a small, fast vessel with four masts. Two of the three ships that set out with Christopher Columbus when he sailed westwards to America in 1492 were caravels. The third ship was a carrack, another very popular type, with three masts and a *forecastle* that hung over the bow. These ships were the forerunners of the great galleons and *ships of the line* of the following centuries. The basic design remained virtually unchanged for nearly 300 years, the only major change being an increase in size.

Exploration by sailors of different

►Christopher Columbus set out on his epic voyage of discovery in 1492. The first European known to have reached the Americas, he made two further voyages there before his death in 1506.

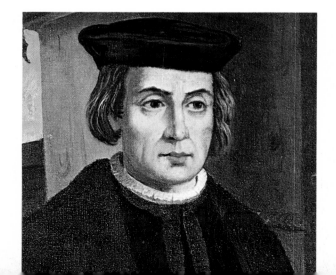

▲A contemporary painting of the battle of Lepanto in 1571 between the Christian and Turkish fleets off the coast of Greece. It was the last time that galleys were used as battleships.

◀The forecastle overhanging the bow, a common feature of 15th and 16th-century warships, shows clearly in this painting of the 16th-century 'Jesus of Lubeck'.

▲A model of the caravel 'Santa Maria', one of the three vessels that made the voyage to America with Columbus in 1492.

Clippers and yachts

During the nineteenth century international trading became fiercely competitive. Faster ships were needed for long ocean voyages. The early steamships were too slow and unreliable. So the fast, sleek sailing ships called clippers were developed. The name clipper came from the fact that they cut down, or clipped, the time taken to sail from one place to another. They were streamlined ships with large areas of sail. The largest had over half a hectare (about 1½ acres) of sail on six or more masts.

Modern sailing vessels include some tall ships, similar in many ways to the clippers. But modern sailing is done largely for pleasure or sport. Most sailing boats range in size from one-man sailboards and dinghies to 25-m (80-ft) ocean-going yachts. The age of sail may not be entirely over, however. In 1980 the Japanese tanker 'Shin-Aitoku-Maru' was launched. Its engines can be assisted by two rigid sails, which are set automatically by a computer. In the future, perhaps, such wind-assisted ships may become more common as fuel becomes more expensive.

nationalities led to disputes over territory and plunder. Soon wars were being fought at sea. The larger ships could carry more cannon and more sail provided more speed. In this way shipbuilders developed the full-rigged fighting ships, such as Sir Francis Drake's 'Revenge' and the much larger 'Victory' used by Sir Horatio Nelson at the battle of Trafalgar in 1805.

◄One of the last of the great sailing ships, the 'Herzogin Cecilie' was built in 1902, but was wrecked off the treacherous south-west coast of Britain in 1936.

FROM STEAM TO SUPERTANKERS

◀Brunel's steamship the 'Great Eastern' was completed in 1858. It was the largest steamship of the time and could carry 4,000 passengers. Steamships did not rely entirely on their engines; the 'Great Eastern' had 5,430 sq m (58,450 sq ft) of sail. In 1866 the 'Great Eastern' succeeded in laying the first transatlantic telegraph cable.

The reign of the clippers lasted only a short time. By the 1860s steamships, which were now much more reliable, were beginning to take over as trading vessels. Steamship owners had one great advantage over their sailing rivals – they could guarantee a date of arrival, regardless of the wind. At first steamships were little more than steam-assisted sailing vessels. And they could not travel far from a land-based supply of fresh water – salt water corroded their boilers. Another problem was that steamships could not carry enough fuel for a long voyage. The early steamships were propelled by paddlewheels, which remained popular for some time, even after the screw propeller was invented in 1836.

The problem of water was solved in 1834 when Samuel Hall, an English engineer, patented a condenser that enabled fresh water to be 'remade' from steam. The problem of fuel was solved by the great engineer Isambard Kingdom Brunel: he simply built larger ships that could carry enough fuel for long voyages. Brunel was the builder of the Great Western Railway and saw steamships as

an extension of his railway across the Atlantic. His first ship, the 'Great Western', sailed from Bristol to New York in 1838. The journey took two weeks. (In 1952 the liner 'United States of America' took 3 days and 10 hours.)

Brunel's ships were built of iron, which lasted much better than wood. In the early twentieth century there was a rapid growth in the construction of warships as Europe prepared for World War 1. These ships needed thick armour plating to withstand the new exploding shells. At the same time the more efficient steam turbine engines replaced the old piston engines. Diesel engines were first used in ships during the 1830s. Today only the largest liners and warships are driven by steam turbines.

Ships have changed little in basic design since the beginning of this century, although new materials have enabled various parts to be refined and improved. Once again there was an increase in size – the liners of the 1930s were four times the *tonnage* of Brunel's largest ship, the 'Great Eastern'.

Today there is a wide variety of types of ship. Submarines were developed by

the Germans in World War 1. Container ships are popular for transporting goods because they can be loaded and unloaded rapidly, as can drive-on/drive-off car ferries. Other specialized ships include hovercraft, aircraft carriers, ice breakers, oil tankers and other bulk carriers. Some ships now have nuclear-powered engines, which do not need refuelling. Nuclear-powered submarines can remain underwater for several months.

▲This hydrofoil is one of the latest types of boat to appear. At speed the foils act like underwater wings to lift the boat clear of the water. Hydrofoils are faster and more economical than conventional boats because less fuel is used up in overcoming water resistance. The surface-piercing V-foils of this hydrofoil act as stabilizers to control rolling and pitching. For example, if the boat tilts to one side, more foil becomes submerged on that side. This creates more lift and the boat is automatically pushed upright.

▼A modern cargo ship. Containerized cargo is stored on deck and in a special hold and can be loaded and unloaded very quickly. Most of the ship is given over to cargo. The crew operates from the small superstructure near the stern.

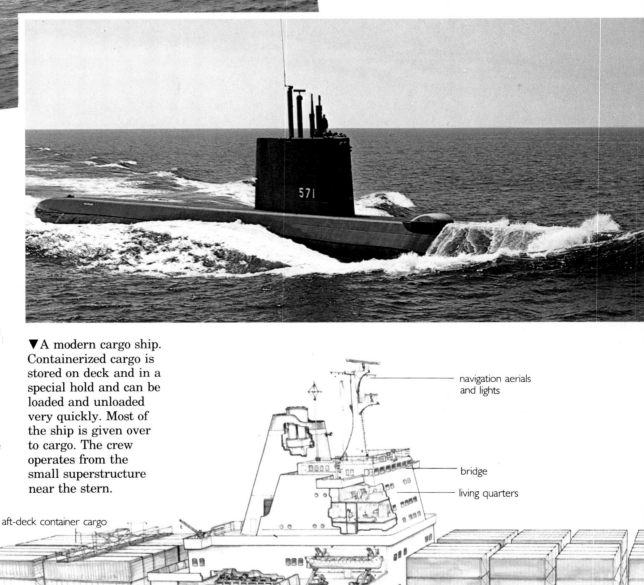

navigation aerials and lights

bridge

living quarters

aft-deck container cargo

loading ramp   rudder   driving screw   engine room   car decks   truck and trailer decks

◀Nuclear power is very useful in icebreakers such as the Russian 'Lenin', which needs an enormous amount of power to break the ice along Russia's northern coast each winter.

▼Luxury liners like the 'Queen Elizabeth 2' or 'QE2' are too slow for long-distance passenger transport, but ideal as cruise ships for tourists.

◀The US submarine 'Nautilus' was the first vessel of any sort to have nuclear propulsion. Nuclear reactors consume no oxygen, so submarines like 'Nautilus' can stay underwater for very long periods. In August 1958 'Nautilus' became the first submarine to travel under the North Pole.

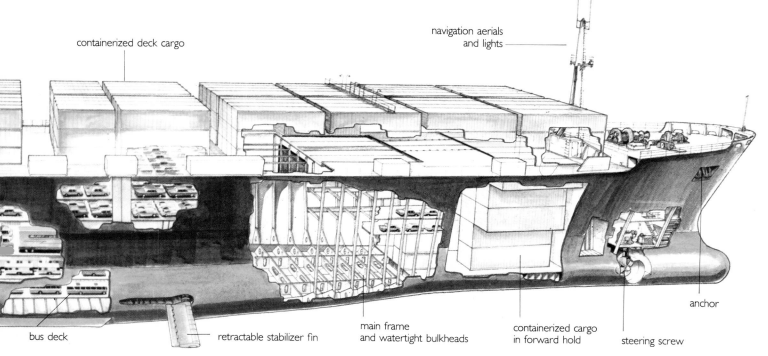

containerized deck cargo

navigation aerials and lights

bus deck

retractable stabilizer fin

main frame and watertight bulkheads

containerized cargo in forward hold

steering screw

anchor

▲The great voyages of exploration spurred the development of navigation into the precise science that it is today. This illustration comes from a book on the subject published in 1583.

▼A portrait of John Harrison holding his chronometer. When he made it in 1761 it was the first to keep time really accurately.

The early sailors used the stars to help them steer in the right direction. At about the time of Christ the Chinese invented the compass. But this instrument did not appear in Europe until about a thousand years later. The mariner's compass, with a magnetic needle attached to a 'floating', pivoted card, appeared in about 1250.

By the fifteenth century sailors had various instruments to help them work out their latitude. These included such devices as the cross-staff, backstaff and astrolabe. All these instruments measured the angle between the horizon and the Sun or a star. The latitude of any point on the Earth's surface is the angular distance, north or south of the equator, measured from the centre of the Earth. In the northern hemisphere this angle is the same as the angle between the horizon and the Pole Star. Thus latitude is easy to measure at night. During the day the angle between the Sun and the horizon is measured at noon. The first really accurate instrument for measuring these angles was the sextant, introduced in 1757.

Finding a ship's longitude was more difficult. The longitude of a point on the Earth's surface is its angular distance

◀Radar is now the chief means of navigation for all boats making long sea voyages. The radar scanners on the foremast of this modern ship are linked to receivers rather like those in the wheelhouse (**below**).

▼The wheelhouse of a modern ship is equipped with a number of navigational aids. The radar receiver on the right shows the direction and distance of other vessels and nearby land. The ship is kept on course by means of a magnetic compass and a gyrocompass. Radio equipment for receiving satellite signals keeps track of the ship's position.

east or west of the Prime Meridian (an imaginary north-south line that passes through Greenwich in London). Because the Earth spins on its axis once every 24 hours, longitude is related to time. This fact was known from the early fifteenth century and over the years many methods of establishing the passing of time aboard ship had been suggested, but the clocks of the period were far too inaccurate. In 1761, however, an English surveyor called John Harrison produced a *chronometer* that lost only 114 seconds during a 147-day journey. With this instrument it was possible to work out a ship's longitude very accurately by comparing local time with the time shown on the chronometer.

Modern navigation depends a great deal upon radio signals. Loran (*Long Range Navigation*) was developed during World War 2. In this system a master station and three slave stations transmit radio signals simultaneously. A receiver on board a ship or airplane calculates the time differences between the signal from the master station and the signals from the two nearest slave stations and uses this information to work out the position of the ship. Many ships and aircraft now use signals from satellites to fix their positions. Other important modern navigational aids include *radar* and *inertial guidance systems*, which use gyroscopes and computers to record every movement of a ship or airplane.

RIVERS AND CANALS

Rivers have been used since earliest times as an effective way of transporting heavy goods, and today navigable rivers are still used to transport goods and passengers. In many cases ports have grown up at the mouths of such rivers, so that goods can be transferred between river vessels and sea-going ships.

Rivers, however, do not always follow the most convenient or desirable routes. And few are navigable for long distances. An artificial canal, on the other hand, can be built where it is most needed. Canals were first built in about 5000 BC in Mesopotamia. They may have developed from the irrigation ditches used to take water to crops.

Locks

The first canals could only pass over flat land. But the invention of the lock meant that canals could be built in areas where the land was not flat. Locks were first developed in China around AD 100. By AD 900 the Chinese had invented the chamber lock, which consists of a chamber of water between two gates. The Chinese used guillotine gates that lifted up and down. The modern mitre gate system was devised by Leonardo da Vinci in 1487. When the gates are closed they form a V-shape that points upstream, the pressure of the water helps to push the gates together and form an effective, water-tight seal. Sluices are used to alter the level of the water in the chamber. When water enters the chamber through the sluices, the level of the water rises, and boats in the chamber are raised. When water is let out through the sluices, the water level is lowered and so are the boats.

Sea-to-sea canals

Several canals have been built with the aim of avoiding the need for long sea voyages. The most famous sea-to-sea canal is the Suez Canal, which links the Mediterranean Sea with the Red Sea. Ships use this canal rather than go around Africa. It was built by the French engineer Ferdinand de Lesseps and completed in 1809. It was enlarged in 1954.

De Lesseps also attempted to make a canal across the Isthmus of Panama in Central America, but he failed and the task was completed in 1914 by the USA. Other canals include the magnificent Corinth Canal in Greece and the Caledonian Ship Canal in Scotland. But as ships have increased in size such canals have become nearly obsolete. Only the Suez Canal is wide enough to take modern vessels.

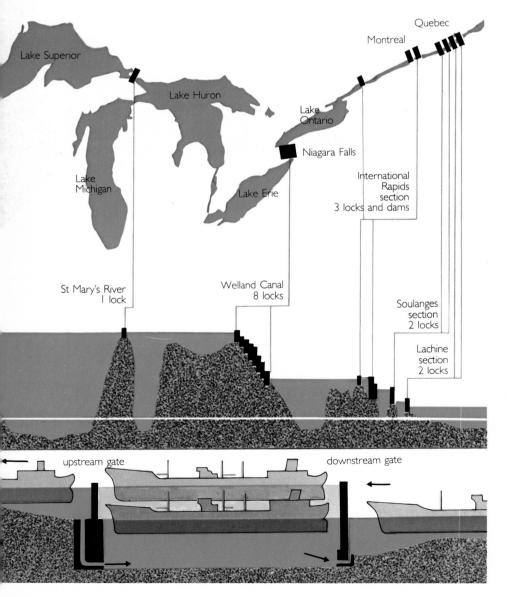

◄The St Lawrence Seaway links the head of Lake Superior with the Atlantic Ocean 3,830 km (2,380 miles) away. Canals and locks are used to raise and lower ships from one lake to the next, bypassing rapids and waterfalls.

◄The River Rhine at St Goarhausen, West Germany. The Rhine is one of the most important European rivers. It runs 1320 km (820 miles) from its source in the Swiss Alps to the port of Rotterdam in the Netherlands and links many German cities with the North Sea. It is also connected to the Black Sea by means of the Rhine-Main-Danube Canal.

►A ship moving through a lock on the Panama Canal with the help of locomotives. The canal was completed in 1914. It is 85 km (53 miles) long and rises 22 m (72 ft) above sea level at its highest point.

▲Five locks close together raise and lower boats up and down a steep slope on the Liverpool-Leeds Canal in England.

The airplane is the form of transport that has probably done most to 'shrink' the world and bring people closer together. Journeys that once took weeks or months can now be completed in a few hours.

BALLOONS AND AIRSHIPS

▼The R.100 airship was built in the late 1920s using duralumin – a very light metal alloy. In 1930 it completed the trip to Canada from England in 78 hours and returned in just 58 hours. However, this was the airship's only trip. After the disastrous crash of its sister airship, the R.101, on 4 October 1930, the R.100 was scrapped.

▶Building a rigid Zeppelin airship in 1935. The engineers used a light metal to build a framework, which was then covered with a fabric 'skin'.

People's early attempts at copying birds and using wings to fly all ended in failure. The first successful method of getting a person into the air was achieved by the two Montgolfier brothers with a hot-air balloon in 1783. After several unmanned flights (including one that took off with a cock, a duck and a sheep!), one of their balloons carried two passengers about 8½ km (5 miles). However, hot-air balloons did not become popular then; gas-filled balloons were used instead. Only recently has hot air

◄Air balloons were the first successful method of air transport for humans. That was in the 18th century, and since then air transport has become much more sophisticated. Today, however, ballooning is becoming a popular sport. It is quiet, causes no pollution and the skies are still uncrowded.

ballooning become a popular sport. Unlike the Montgolfier balloons, a modern hot-air balloon can carry its own heat source – a gas burner fuelled by bottled gas.

The first gas-filled balloon, devised by the French scientist Professor Jacques Charles, also appeared in 1783. In 1785 a balloon carrying Jean-Pierre Blanchard and Dr John Jeffries crossed the English Channel in 2½ hours. Nine years later, in 1794, a balloon was used in war for the first time, when Captain Coutell of the French army observed the positions and movements of the Austrian army before the battle of Fleurus. The French won the battle. Gas-filled balloons are still used, largely for scientific purposes.

In 1852 a gas-filled balloon powered by a steam engine made its first flight. This machine, built by the French engineer Henry Gifford, was therefore the first airship. During the next 35 years many

airships were built – some were powered by steam engines, others by electric motors. But the breakthrough came with the invention of the internal combustion engine and the development of lightweight metals with which to build them. In the late 1890s the first rigid airships began to appear and by the mid-1930s many such airships had been built in Germany, Britain and America.

Most of the rigid airships were filled with the highly flammable gas hydrogen. Several disastrous accidents, such as the one that destroyed the 'Hindenberg' in 1937, virtually put an end to airship development for many years. Recently, however, airships have begun to appear again, using the much safer gas helium. Modern airships may become very useful. They do not travel as fast as an airplane, but they use less fuel, can carry greater loads and do not need large take-off areas.

AIRPLANES

▲Because flying was a totally new form of transport, the designers of the first aircraft had no models to copy, so many of the early airplanes were rather odd-looking machines.

▼The Wright brothers made their first successful flights in 1903. Five years later, in 1908, they brought one of their planes to Europe (by boat). Here Wilbur gives a demonstration flight outside Rome.

The first airplanes to succeed in getting off the ground were gliders developed by the British inventor George Cayley between 1804 and 1859. The German pioneer Otto Lilienthal built eighteen gliders and made thousands of flights between 1891 and 1896, when he died as the result of a crash.

Early powered flight

By this time the search for powered aircraft was on. Many weird and wonderful machines were tried. In 1890 Clement Ader, a Frenchman, achieved a powered 'hop' in his bat-winged, steam-powered *monoplane* 'Eole'. Finally, on 17 December 1903 the American Orville Wright, helped by his brother Wilbur,

▲One of Sir George Cayley's triplane gliders. Cayley never flew in any of his gliders. Early models were flown with a boy on board and in 1859 Cayley launched his coachman, who resigned his job immediately afterwards.

▶In August 1909 the French held an aviation (flying) week at Rheims. This is the poster advertising what proved to be the first large international flying meeting. The era of the airplane had begun.

made the first sustained powered flight in their Wright Flyer I.

By 1905 the Wright brothers had achieved a flight of 39 km (24 miles). But few others were having much success. The first powered flight in Europe was not made until 1906 when Alberto Santos-Dumont, a Brazilian who had been a pioneer of airships, flew 220 m (720 ft) in his Santos-Dumont 14-*bis*. Soon others, such as Louis Blériot, Gabriel Voisin and Henri Farman were also getting off the ground. Voisin and Farman built successful box-kite airplanes and the first fully controlled flight in Europe was made by the Voisin-

200.000F. de PRIX
GRANDE SEMAINE
D'AVIATION DE LA
REIMS DU 22 AU 29 AOÛT 1909 CHAMPAGNE
E. Montaut
Affiches MONTAUT & MABILEAU 84. Rue d'Amsterdam, PARIS.

Farman I in 1907. In 1908 Wilbur Wright brought a Wright Flyer IV to Europe and amazed everyone by making two-hour flights. Farman saw that the Wrights had much greater control over their plane than anything yet achieved in Europe. Inspired by this he added ailerons to his aircraft. These provided much greater

control of the aircraft and he succeeded in flying 27 km (16 miles). In 1909 Louis Blériot flew across the English Channel in a Blériot XI, one of his highly successful monoplanes.

From racers to fighters

By 1910 aircraft companies were springing up rapidly. At this stage aircraft were mostly being used in racing competitions and speed trials. Aircraft and their engines were improving rapidly and speed records were constantly being broken. Among the many notable aircraft of the time were the seaplanes built by Glenn Curtiss in the USA. In 1912 the British designer Alliot V. Roe built the Avro F, the first aircraft to have an enclosed cockpit and cabin. In Russia, the first four-engined aircraft, the 'Russkii Vitiaz' built by Igor Sikorsky, flew in 1913.

►Louis Blériot built a number of monoplanes in the early 1900s. In July 1909 he was the first to fly across the Channel from France to England.

▼The Sopwith Pup was a fast, easy-to-handle fighter plane of World War 1. It carried one machine gun and could fly at speeds of up to 180 km/h (111 mph).

By the time World War 1 broke out, the military value of aircraft had been realized. To begin with they were used only for reconnaissance, but within a year the battle for supremacy in the air was on. In 1915 the French Morane-Saulnier L became the first airplane to be equipped with a machine gun. The German Fokker E III, or Eindekker, introduced in the same year, had a forward firing machine gun that fired between the rotating blades of the propeller. By the end of the war both sides had a number of fighter aircraft and much larger bombers.

Between the wars

The impetus that World War 1 gave to aircraft manufacture continued after the war. All kinds of aircraft were built, including trainers, civil transporters, racers and small private aircraft. In 1919

John Alcock and Arthur Whitten Brown made the first non-stop crossing of the Atlantic in a Vickers Vimy.

The first large passenger aircraft were developed and airline services began. By the 1930s aircraft were carrying passengers to many parts of the world. Flying boats were popular for long journeys across oceans and wherever there were few suitable airfields. But there were also superb land aircraft, such

▲The Ryan NYP aircraft, named the 'Spirit of St Louis', was the aircraft in which Charles Lindberg made the first solo non-stop crossing of the Atlantic Ocean in 1927.

▼The Junkers Ju 52 was a popular transport plane just before World War 2.

as the Handley Page H.P. 42s used by Imperial Airways. These aircraft were only superseded by the new all-metal, low-wing monoplanes with retractable undercarriages, such as the Boeing 247 (1933) and the Douglas DC-2 (1934). These sleek aircraft marked the start of a new era both in flying and travel.

▼The Supermarine Spitfire was one of the most famous fighters of World War 2.

World War 2

Many now legendary aircraft were designed and built during World War 2. The *biplane* fighters of World War 1 were replaced by fast, monoplanes, such as the German Messerschmitt Bf 109 (popularly known as the Me 109) and the British Hawker Hurricane and Supermarine Spitfire. The British fighters were guided to their targets by radar. Later in the war the Bristol Beaufighter was equipped with radar for intercepting enemy aircraft at night. Other famous World War 2 fighters included the British de Havilland Mosquito, Germany's best fighter the Fockewulf Fw 190, the American Mustang and the Japanese Mitsubishi A6M Reisen, or Zero, which was used in the attack on Pearl Harbor.

German bombers included the Junkers Ju 88, which was probably the most versatile aircraft of the Luftwaffe (the German air force). British bombers included the Vickers Wellington and the famous Avro Lancaster – the aircraft that took part in the 'Dam Busters' raid against the Moehne Ader dam in Germany in May 1943. The American Boeing B-17 Flying Fortress and the Consolidated B-24 Liberator were among the largest of the World War 2 bombers.

▼The legendary Boeing B-17 Flying Fortress bomber was used by the Americans for daylight bombing raids during World War 2.

THE JET AGE

The first jets

The jet engine was developed during World War 2 in both Germany and Britain. The first person to propose a practical scheme for such an engine was the British engineer Sir Frank Whittle in 1928. He tried out his first engine in 1937, but few people were interested in it.

Meanwhile, in Germany Dr Hans von Ohain was also working on jet engines and had the good fortune to interest Ernst Heinkel. The result was that in 1939 the Heinkel He 178 was the first jet aircraft to fly. Further development led to the production of the Messerschmitt 262A, which in 1944 was the first jet fighter to be used in action.

In Britain the result of Whittle's work was the Gloster E28/29, which flew in 1941. In 1945 the first British jet fighter was the Gloster Meteor. After the war development of jet aircraft continued rapidly. The first jet airliner, the de Havilland D.H. 106 Comet I, appeared in 1952.

Jet engines

A jet engine works by creating a backward-flowing stream of hot gases that pushes against the air. At the front of the engine air is drawn in and compressed into a smaller space by the rotating blades of a compressor. The compressed air is fed into combustion chambers where fuel is added. When the fuel/air mixture is ignited, the expanding hot gases are forced out of a nozzle at the rear of the engine. In the process they turn a turbine, which drives the compressor. This is the basic jet engine, or turbojet, which is used in most high speed aircraft.

The turboprop engine is a variation of the basic turbojet. It works in exactly the same way, but the turbine also drives a propeller, which helps to pull the aircraft along. Turboprop engines work better at low speeds than turbojets. The first turboprop airliner was the Vickers-Armstrong Viscount, which appeared in 1953. However, turboprops have not been used in large aircraft since 1955.

Between high speed turbojets and low speed turboprops comes a third group of engines known as turbofans. In this type of engine a turbine drives a fan at the front. The fan acts as a large compressor and the air it compresses is divided into two parts. One part enters the main compressor, the other part flows round the outside of the combustion chamber, helping to cool the engine and reduce noise. Turbofan engines produce much greater power at lower working speeds than turbojets and are much quieter. Today they are the most widely used

▼The basic type of jet engine is the turbojet, in which the exhaust gases turn a turbine, which turns the compressor. In a turboprop the turbine also turns a propellor. In a turbofan the turbine turns a low-pressure compressor, or fan. The turbofan shown here has afterburners; fuel is burned in the hot exhaust gases to provide extra thrust when needed, for example at take off.

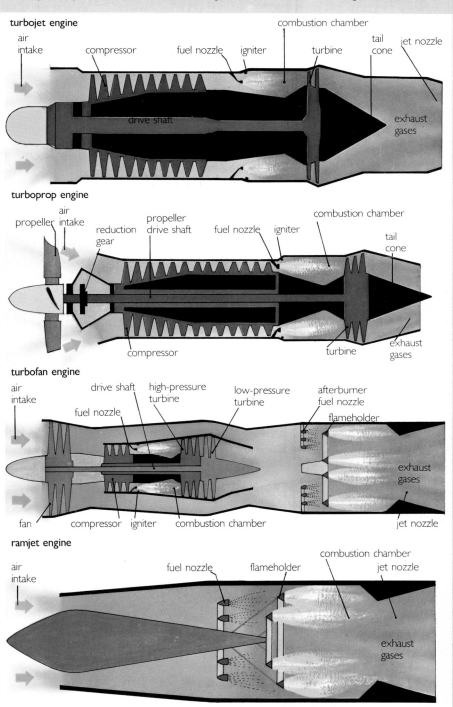

turbojet engine
air intake — compressor — fuel nozzle — igniter — combustion chamber — turbine — tail cone — jet nozzle — drive shaft — exhaust gases

turboprop engine
propeller — air intake — reduction gear — propeller drive shaft — fuel nozzle — igniter — combustion chamber — tail cone — compressor — turbine — exhaust gases

turbofan engine
air intake — drive shaft — high-pressure turbine — low-pressure turbine — afterburner fuel nozzle — flameholder — fuel nozzle — exhaust gases — fan — compressor — igniter — combustion chamber — jet nozzle

ramjet engine
air intake — fuel nozzle — flameholder — combustion chamber — jet nozzle — exhaust gases

▲The de Havilland Comet was the first airliner powered by turbojets.

engines in large aircraft. Powerful turbofan engines are used in modern jumbo jets, such as the Boeing 747.

Supersonic flight

The development of the jet engine meant that aircraft speeds increased dramatically after World War 2. In 1954 an airplane flew faster than sound for the first time. This was the American F.100 Super Sabre, which could reach speeds of up to 1350 km/h (850 mph) – sound travels at about 1220 km/h (760 mph). Today the official air speed record is held by a Lockheed SR71A, a reconnaissance jet aircraft, which flew at just over 3,529 km/h (2,193 mph) in 1976. Rocket-powered aircraft have flown even faster.

▼The Lockheed SR-71 is one of the world's most advanced aircraft. Nicknamed the 'Blackbird', it can reach speeds more than three times the speed of sound.

◄The General Dynamics F-111 fighter-bomber has swing wings. At low speeds they are held out sideways, but in supersonic flight they are moved to the swept-back position to reduce drag. The F-111 can fly at over 2,400 km/h (1,490 mph).

The speed of a supersonic aircraft is described in Mach numbers. Mach 1 is the speed of sound, Mach 2 is twice the speed of sound, and so on. When an airplane is flying below Mach 1, the pressure waves caused as it disturbs the air through which it is flying travel ahead of it at the speed of sound. But at Mach 1 the pressure waves build up, creating a shock wave in front of the aircraft. At speeds above Mach 1 the pressure waves are left behind and the shock wave forms a cone. People on the ground below hear a 'sonic boom' as the shock wave passes.

Above about 800 km/h (500 mph) aircraft wings that stick straight out sideways tend to cause excessive drag and this becomes much worse as the aircraft approaches Mach 1. To help overcome this problem supersonic aircraft have

▲ The supersonic airliner Concorde was built by Britain and France. It was introduced into service in 1976.

thin, swept-back wings. Aircraft that travel at over 2,250 km/h (1,400 mph) have special delta (triangular) wings. Some supersonic aircraft have swing wings. These are held straight out sideways at low speeds and then swung into the swept-back position for supersonic flight.

Today's supersonic aircraft are mostly fighters and other military aircraft. The only supersonic passenger aircraft that has been built so far is Concorde. Future supersonic transports (SSTs) will have to be less noisy and more economical to operate before they are widely used.

TAKING-OFF VERTICALLY

Most aircraft need a fairly long runway to take off. But since the earliest days of flight people have tried to devise machines that take off vertically. During the 1840s George Cayley built models of 'rotating wing' aircraft. Even when fixed-wing airplanes had been shown to work well, there were those who remained convinced that such machines were too dangerous.

In 1907 two machines did manage to get off the ground, but only for a few minutes. By 1924, five years after Alcock and Brown had crossed the Atlantic in a fixed-wing aircraft, the distance record for a helicopter was still only 800 m (2,624½ ft).

Autogyros and helicopters
The breakthrough came with the invention and development of the autogyro by the Spaniard Juan de la Cierva between 1919 and 1936. He added a freewheeling rotor to an ordinary fixed-wing airplane. When the aircraft was moving forward, the rotor automatically turned in the airflow and as it turned it produced extra lift. An autogyro cannot take off vertically, but it requires only a very short runway.

By the 1930s a number of autogyros had been produced. And as a result of experience with autogyro rotors, some engineers were beginning to have more success with helicopters. The first

▼Helicopters are used for a wide range of tasks. The Sikorsky S-64 Skycrane shown here is designed for lifting heavy loads. If required a cargo pod can be fitted to the helicopter behind the cockpit.

practical machine was the Focke-Achgelis
Fa-61, which flew for the first time in
1936. It had two rotors that turned in
opposite directions. This was to overcome
the problem of torque – the tendency for
a helicopter to spin in the opposite
direction to its rotor. In 1939 in the USA
Igor Sikorsky solved the problem in a
different way. He used one horizontal
rotor together with a vertical rotor on the
tail. By 1942 Sikorsky's VS-316A was the
first practical single rotor helicopter.
Today, helicopters come in a tremendous
range of sizes and designs and some are
even powered by jet engines.

Jump jets

The possibility of using jet engines for
vertical take-off was investigated in the
1950s. The first experimental craft was
built by the Rolls-Royce company and
nicknamed the 'Flying Bedstead'. It had
four legs and a framework of tubular

steel surrounding two Rolls-Royce Nene
turbojets. The 'pilot' sat on top and
controlled four downward-pointing
nozzles. The success of this strange
machine led to the production of a
special, lightweight engine, the RB 108.
Short Brothers & Harland used five of
these in their SC-1 research aircraft –
four for lift and one for propulsion. In
1958 the SC-1 became the first fixed-wing
aircraft to achieve vertical take off.

Since then various systems have been
used, including propellers mounted on
wings that tilt and tilting jet engines.
But the most effective system has proved
to be the one known as vectored thrust.
In this system the jet engines are
mounted horizontally and remain fixed.
But the exhaust gases can be directed
downwards or backwards as required by
nozzles that swivel. This system is the
one used in the world's most versatile
aircraft, the Hawker Siddeley Harrier.

▲A Hawker Siddeley
Harrier takes off. Its
nozzles direct the blast
from its two jet engines
vertically downwards
to lift the aircraft off
the ground. To fly
forwards the nozzles
are swivelled to point
rearwards.

AIRCRAFT NAVIGATION

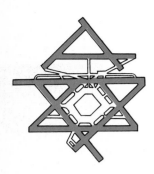

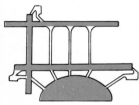

▲Aircraft take off and land into the wind. At airports where the wind direction varies a lot, there are a number of runways (top and centre). Where the prevailing wind is more or less constant there are only a few runways (bottom).

►At a busy airport aircraft waiting to land fly in a holding pattern, or 'stack', round a radio beacon. When an aircraft leaves the bottom of the stack to land it flies to the start of the approach route. There the instrument landing system picks up the radio signals from aerials on the ground. Information about the aircraft's position is displayed on cockpit instruments so that the pilot can keep the aircraft on the correct glide path.

In the early days of flying pilots had no navigational aids. They had to find their way about by picking out recognizable landmarks. Following railway lines was often a good way of finding places and sometimes the name of the railway station was written in large white letters on the roof to help pilots.

It was soon discovered that an aircraft does not necessarily fly in the direction in which it appears to be heading. Wind can cause an airplane to drift several degrees from its original course. Early navigators used a process known as *dead reckoning* to work out how to fly in a particular direction. Today there are electronic instruments for determining the rate of drift.

Cockpit instruments

The magnetic compass was and still is an important navigational instrument. The marine sextant was too awkward to use in an aircraft and a slightly less accurate instrument called a bubble sextant was devised instead. This measures the angle between the Sun and the horizon and because it has an artificial horizon it can be used above the clouds. An altimeter tells a pilot how far off the ground the

aircraft is and an airspeed indicator shows how fast the plane is moving through the air. An artificial horizon indicates the aircraft's altitude (the angle at which the wings are tilted from the horizontal). Modern aircraft have a wide variety of instruments, including a rate of climb/descent indicator and a turn/slip indicator, which shows the rate at which the plane is turning.

Gyroscopes are used a great deal in modern aircraft. The artificial horizon contains a gyroscope and a gyrocompass is set so that it always points towards true North. Autopilots and inertial guidance systems use gyroscopes to detect the movements of an airplane so that any corrections needed to keep it on course can be made automatically.

Radio direction finding

As in ships, radio now plays an important part in aircraft navigation. Early types of radio direction-finding (D/F) systems used two ground stations. An airplane was equipped with an aerial formed from a loop of wire. This could be rotated until the signal from one station faded out completely. This gave a bearing on the station. Then a bearing on the second

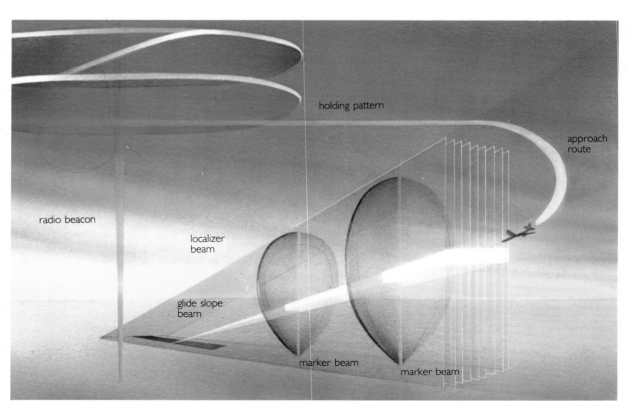

holding pattern

approach route

radio beacon

localizer beam

glide slope beam

marker beam

marker beam

station was taken in the same way and the two bearings gave a 'fix' on the aircraft's position.

Later systems used more stations to give greater accuracy. In 1940 British engineers devised the Gee system, which used three stations to help guide allied bombers to their targets. In 1945 the Decca system which uses four stations was introduced, followed quickly by the Loran system.

Airports

Airports are very busy places and it is vital to keep aircraft well clear of each other. Air traffic controllers follow the movement of aircraft on radar screens and instruct each pilot on any changes of course and height that may be necessary. In the busy airspace around an international airport pilots navigate using radio beacons on the ground. Radio beacons also mark the stacking areas, at which aircraft arriving at peak times are held before their turn comes to land.

Many airports now use instrument landing systems (ILS). Signals from two aerials are picked up by instruments in the aircraft and the pilot can see if the aircraft moves away from the correct glide path. If the ILS is linked to the aircraft's autopilot and other instruments, such as radio altimeters, the aircraft can be landed automatically.

Airports, particularly the major international ones, have to be designed very carefully to allow for the swift movement of large numbers of people and planes and great quantities of luggage and cargo. The aerial photograph (**left**) shows the layout of New York's Kennedy Airport. The TWA terminal building there is shown **above**.

6 THE FUTURE

Communications

It is already possible to see how the communications industry is likely to progress in the future. The basis of tomorrow's communication system will be the telephone, the television and, above all, the computer.

Telephones and computers can already be linked (page 00). But at present an intermediate decoding device, or modem, is needed. As more and more electronic telephone exchanges are introduced, modems will become unnecessary as telephones and computers will use the same language of digital electric pulses. It will therefore become much easier and quicker to communicate with computers by telephone.

Televisions, known in the computer world as visual display units (VDUs), are already a common and useful way of displaying computer information. If telephones are now linked with televisions to produce videophones, callers will, if they wish, be able to see as well as talk to each other. This may or may not become popular, but it is now possible to forsee a worldwide communications system, by which people can communicate with each other, information can be transferred between computers and people can communicate with computers very quickly and easily.

Such a system will be useful to all kinds of businesses, including manufacturers, exporters and importers, insurance companies and banks. Many more people will work at home, using video/computer consoles to keep in touch with their offices and colleagues. In the home it will be possible to use such a system for routine shopping, simply by calling up the computer of the local supermarket.

Transport

Computers and other electronic devices will also play an increasingly important part in transport. The first completely robot cars will probably be autotaxis in cities, where there could also be moving walkways.

Outside cities there will still be cars and lorries, but these will probably be highly computerized, streamlined vehicles. As oil-based fuels become more expensive, other forms of fuel, such as natural gas and alcohol, may become more widely used.

Trains of the future may be pulled by steam locomotives. But these will bear no resemblance to the old types of locomotive. They will produce steam by burning powdered coal in specially designed furnaces. The steam will not be released, instead it will be condensed and reused. Efficient burning of the fuel will mean that these locomotives will be much cleaner than the old type.

Other forms of train may have no wheels. Instead they will be held off their tracks by magnetic levitation and propelled by linear electric motors or gas turbines. These maglev trains could run on monorails or they could run in specially designed tunnels from which all the air is removed. Such vacuum bullet trains could travel incredibly fast.

Ships of the future may be larger and more dependent on electronics. Huge container ships – possibly even cargo submarines – may appear. Satellite navigation may be developed to the point where satellites relay course and speed instructions from land-based controllers.

It is possible, but unlikely, that planes will become much larger. On the other hand airships may become much more common. One of the latest forms of transport, the space shuttle, will become increasingly useful if space stations are built. A similar form of transport could be used to carry people and goods rapidly between distant places on the Earth's surface. A spaceplane would take off like a rocket, level out above the atmosphere and then glide down to its destination.

▼Commander Vance Brand (left) and pilot Bob Overmyer watch the fiery glow made as 'Columbia' re-enters the Earth's atmosphere at about Mach 15 (15 times the speed of sound). Perhaps in the future such space journeys will become more common and not be limited to a few astronauts.

GLOSSARY

Articulated locomotive A steam locomotive with two sets of cylinders, each of which drives a separate set of wheels. The frames supported by the wheels are hinged together. In Garratt locomotives, a short wide boiler is slung between the two engine units. A Mallet locomotive has a long, narrow boiler supported over the top of the two units.

Axle A metal or wooden shaft linked to a wheel or pair of wheels. If the axle is fixed to the wheel, it turns as the wheel turns. Alternatively, the axle may be fixed to the vehicle, in which case the wheel turns freely on the axle.

Beam bridge A bridge consisting of a single, horizontal beam across the space being bridged. A long beam bridge may be supported from underneath by one or more uprights.

Biplane An airplane with two pairs of main wings, one pair above the other.

Bogie A chassis and set of wheels on a railway locomotive or carriage that can move independently of other sets of wheels. A railway carriage, for example, usually has two bogies, one at each end. The bogies are able to pivot under the carriage, so the carriage can negotiate fairly sharp bends.

Box girder bridge A type of beam bridge in which the beam is made up of one or more hollow, rectangular tubes built of iron girders and plates. Railway lines may be laid inside the tubes.

Cantilever bridge A bridge with two main supports, between which is a long span. The ends of the bridge are fixed firmly to the bank and this helps carry the load in the middle of the long span.

Cathode ray tube An electronic device basically consisting of a glass tube containing a vacuum and two metal electrodes – a negative cathode and a positive anode. When the electrodes are connected to a high voltage source of electricity, electrons stream from the cathode to the anode. Using suitable deflecting devices, the electrons can be aimed at a fluorescent screen on the wall of the glass tube. This glows when struck by electrons.

Chronometer A device for measuring time, such as a clock or a watch.

Crystal A solid material that has a definite geometrical shape, eg salt, diamond. The crystals used in early radio sets included those of silicon, galena (lead sulphide) and carborundum (silicon carbide).

Cuneiform This word means wedge-shaped. It is used to describe the wedge-shaped characters that make up the writing of the ancient peoples of Mesopotamia and Persia.

Dead reckoning 1. A system of navigation in which the position of a ship or aircraft is calculated by determining how far and in what direction the craft has travelled since its last known position. 2. A system of aircraft navigation in which a pilot, knowing the speed and direction of the wind, calculates the heading (the direction in which the aircraft points) necessary for the aircraft to follow a particular flight path (the actual course of the aircraft).

Forecastle The part of a ship near the bow, often abbreviated to fo'c'sle. The term dates from the fifteenth century when, in order to raise archers and other fighting men above the decks of ships belonging to enemies, shipbuilders constructed 'castles' at each end of a ship (forecastle and after-castle).

Geostationary orbit A satellite orbit that follows the line of the equator 35,900 km (22,307 miles) above the Earth's surface in the same direction in which the Earth spins. At this height the satellite moves at the same rate as the Earth spins and therefore always remains above the same point on the equator.

Hieroglyphics A form of writing in which picture symbols are used to represent ideas and objects.

Industrial Revolution The historical period when factories and machines came into widespread use for the first time.

Inertial guidance system A system in

which gyroscopes are used to detect all the movements of a ship or aircraft – starting, stopping, speeding up, slowing down or changing direction. The information from the gyroscopes is fed to a computer, which can then work out how far and in what direction the ship or aircraft has travelled.

Internal combustion engine An engine in which combustion (burning) of the fuel takes place inside the engine, usually in special cylinders. Petrol, gasoline and diesel engines are internal combustion engines, whereas in steam engines the fuel is burned outside the cylinders and boiler.

Ionosphere A layer in the upper part of the Earth's atmosphere that reflects long, medium and short wave transmissions. VHF (very high frequency) and UHF (ultra-high frequency) transmissions pass through this layer. The layer is so-called because the atoms of gas it contains are ionized; that is, turned into electrically charged atoms, or ions, by radiation from space.

Laminated spring suspension Method of suspending a vehicle on one or more of its axles using springs made up of layers or leaves, of steel plate. The leaves are held together by metal clips and each one overlaps the one below it at both ends. The axle is attached at the centre, where the layers of leaf springs are thickest and strongest. The ends of the longest leaf spring are attached to the chassis.

Latitude The latitude of a point on the surface of the Earth is its angular distance north or south of the equator measured from the centre of the Earth. The equator lies at 0° latitude, the North Pole is at 90°N and the South Pole is at 90°S. Any line on the Earth's surface parallel to the equator is a line of latitude.

Longitude The longitude of a point on the Earth's surface is its angular distance east or west of the Prime Meridian (0° longitude). The easiest way of measuring longitude is by time and 1 hour of time is equal to 15° of longitude. The Prime Meridian passes through a point at the Royal Observatory at Greenwich, London. This meridian was agreed internationally at a congress in 1884.

Middle Ages The period of European history from the end of the Roman empire, about AD 500, to the start of the Renaissance in the mid-fifteenth century.

Monoplane An airplane with one pair of main wings.

Pantograph A device attached to the top of an electric locomotive that provides the link between the locomotive and the overhead wires that supply it with electric current. The simplest form consists of two diamond frames with copper or carbon contact strips at the top. The frames are pivoted together on the locomotive and springs keep the contacts touching the wires.

Radar *Ra*dio *d*etection *a*nd *r*anging. A system that uses radio waves to detect the position of objects. The radio waves are sent out by a transmitter/receiver via a rotating aerial. The waves that are reflected by objects return to the aerial. In the receiver the distance and direction of each object is calculated using the time difference between sending and receiving the signal and the position of the aerial. An electrical signal containing this information is sent to a device that works rather like a television and the objects appear as luminous 'blips' on a screen.

Ship-of-the-line Sailing warships of the eighteenth and nineteenth centuries.

Telegraph A system of communication in which messages can be sent over a long distance. Optical telegraph uses semaphore towers or flashing lamps. Electrical telegraph uses wires to send coded electrical signals – for example the 'dots' and 'dashes' of the Morse Code. Modern telegraph systems include telex and TWX. Teletex is a modern high speed system designed to link word processors.

Telephone A system of communication in which speech is carried along wires in the form of electrical signals.

Tonnage 1. The weight of an object in tons (1 ton = 2240 lb = 1016 kg) or tonnes (1 tonne = 1000 kg). 2. A measurement of the size of a ship, based on the weight of water displaced by the unladen ship (1 ton of seawater occupies 35 cubic feet; 1 tonne of seawater occupies 1 cubic meter).

Transistor An electronic device, with no moving parts, that can be used as a switch or to amplify an electric signal.

INDEX

Acknowledgments

Aerofilms, Air Portraits, Aldus Archive, George Allen and Unwin/ Doubleday, Bang and Olufsen, R Bastin, Bibliotheque Nationale Paris, A Bollo, British Airways, British Leyland, Charles Brown, Camera Press, AC Cooper, G Costa, Jim Cutler, Dalgi Orti, Ian Dobbie, Flight Picture Library, GEC General Signal Ltd, Michael Holford, Robert Hunt Library, Archivio IGDA, The Image Bank, Kobal Collection, Lancia Public Relations, Lufthansa, MARS/ Deutsche Bundesbohm/ Adler, Mansell Collection, Masters and Fellows of Magdalene College Cambridge, McAlpine, JG Moore, NASA, National Maritime Museum, National Motor Museum, Novosti, Picturepoint, The Photo Source, T Poggio, Popperfoto, C Pozzoni, Archivio Radaelli, A Rizzi, Rodriguez, GA Rossi, Science Museum London, Ronald Sheridan, Skyphotos, Spectrum Colour Library, Terence Spencer, YRM Architects and Planners.